T0320459

Air Pollution and the Electromagnetic Phenomena as Incitants

Electromagnetic frequencies are everywhere in our daily lives. This brand-new series on EMF describes how having an understanding of the vast combinations of electrical and chemical problems will help in the diagnosis and treatment of electromagnetic sensitivities. This series covers the work of a renowned scientist in the field whose interest ranges from reversing the dysfunction of chronic illness to optimizing the health of his patients.

Air Pollution and the Electromagnetic Phenomena as Incitants is the first in an extensive series written by William J. Rea on electromagnetic sensitivity and the impact of electromagnetic phenomena on our lives. The complete list of books soon to be part of the series is as follows:

Air Pollution and the Electromagnetic Phenomena as Incitants
The Physiological Basis of Homeostasis for EMF Sensitivity –
Molds, Foods and Chemicals
Pollutant Entry and the Body's Homeostatic Response to and
Fate of the Noxious Stimuli
EMF Effects from Power Sources and Fixed Specialized
Equipment; Electro Smog
(Dirty Electricity) from Communication Equipment
Basic Science and Science of Clinical Electromagnetic Sensitivity
Physiology of Electrohypersensitivity; EMF Treatment

Air Pollution and the Electromagnetic Phenomena as Incitants

Part of the Electromagnetic Frequency Sensitivity Series

William J. Rea
Kalpana D. Patel

CRC Press
Taylor & Francis Group
Boca Raton London New York

CRC Press is an imprint of the
Taylor & Francis Group, an **informa** business

CRC Press
Taylor & Francis Group
6000 Broken Sound Parkway NW, Suite 300
Boca Raton, FL 33487-2742

Printed on acid-free paper

International Standard Book Number-13: 978-0-367-00054-7 (Hardback)

Visit the Taylor & Francis Web site at
http://www.taylorandfrancis.com

and the CRC Press Web site at
http://www.crcpress.com

Contents

Authors

William J. Rea, M.D. is a thoracic, cardiovascular, and general surgeon with an added interest in the environmental aspects of health and disease. Dr. Rea founded the Environmental Health Center-Dallas (EHC-D) in 1974 and is currently director of this highly specialized Dallas-based medical facility.

Dr. Rea was awarded the Jonathan Forman Gold Medal Award in 1987 for his exceptional research in environmental medicine, The Herbert J. Rinkle Award in 1993 for his outstanding teaching, and the 1998 Service Award, all by the American Academy of Environmental Medicine. He was named Outstanding Alumnus by Otterbein College in 1991. Other awards include the Mountain Valley Water Hall of Fame in 1987 for research in water and health, the Special Achievement Award by Otterbein College in 1991, the Distinguished Pioneers in Alternative Medicine Award by the Foundation for the Advancement of Innovative Medicine Education Fund in 1994, the Gold Star Award by the International Biographical Center in 1997, Five Hundred Leaders of Influence Award in 1997, Who's Who in the South and Southwest in 1997, The Twentieth Century Award for Achievement in 1997, the Dor W. Brown, Jr., M.D. Lectureship Award by the Pan American Allergy Society, and the O. Spurgeon English Humanitarian Award by Temple University in 2002. He has authored 10 medical textbooks, including *Chemical Sensitivity (V. 1–4)* and *Reversibility of Chronic Degenerative Disease and Hypersensitivity, V. 1: Regulating Mechanisms of Chemical Sensitivity*, and is co-author of *Your Home, Your Health and Well-Being.* He also published the popular "how to" book on building less polluted homes, *Optimum Environments for Optimum Health and Creativity.* Dr. Rea has published more than 150 peer-reviewed research papers related to the topic of thoracic and cardiovascular surgery as well as that of environmental medicine.

Dr. Rea currently serves on the board and is president of the American Environmental Health Foundation. He is vice president of the American Board of Environmental Medicine and previously served on the board of the American Academy of Environmental Medicine. He previously held the position of chief of surgery at Brookhaven Medical Center and chief of cardiovascular surgery at Dallas Veteran's Hospital, and he is a past president of the American Academy of Environmental Medicine and the Pan American Allergy Society. He has also served on the Science Advisory Board for the U.S. Environmental Protection Agency, on the Research Committee for the American Academy of Otolaryngic Allergy, and on the Committee on Aspects of Cardiovascular, Endocrine and Autoimmune Diseases of the American College of Allergists, Committee on Immunotoxicology for the Office of Technology Assessment, and on the panel on Chemical Sensitivity of the National Academy of Sciences. He was previously adjunct professor with the University of Oklahoma Health Science Center College of Public Health. Dr. Rea is a fellow of the American College of Surgeons, the American Academy of Environmental Medicine, the American College of Allergists, the American College of Preventive Medicine, the American College of Nutrition, and the Royal Society of Medicine.

Born in Jefferson, Ohio and raised in Woodville, Ohio, Dr. Rea graduated from Otterbein College in Westerville, Ohio, and Ohio State University College of Medicine in Columbus, Ohio. He then completed a rotating internship at Parkland Memorial Hospital in Dallas, Texas. He held a general surgery residency from 1963 to 1967 and a cardiovascular surgery fellowship and residency from 1967 to 1969 with The University of Texas Southwestern Medical

School system, which includes Parkland Memorial Hospital, Baylor Medical Center, Veteran's Hospital, and Children's Medical Center. He was also part of the team that treated Governor Connelly when President Kennedy was assassinated.

From 1969 to 1972, Dr. Rea was assistant professor of cardiovascular surgery at the University of Texas Southwestern Medical School; from 1984 to 1985, Dr. Rea held the position of adjunct professor of environmental sciences and mathematics at the University of Texas, while from 1972 to 1982, he acted as clinical associate professor of thoracic surgery at The University of Texas Southwestern Medical School. Dr. Rea held the First World Professorial Chair of Environmental Medicine at the University of Surrey, Guildford, England, from 1988 to 1998. He also served as adjunct professor of psychology and guest lecturer at North Texas State University.

Kalpana D. Patel, M.D. is a pediatrician with an added interest in the environmental aspects of health and disease. Dr. Patel founded the Environmental Health Center-Buffalo (EHC-Buffalo) in 1985, a specialized Buffalo-based medical facility. Dr. Patel was awarded the Jonathan Forman Gold Medal Award in 2006 for outstanding research in environmental medicine and the Herbert J. Rinkle Award in 2008 for outstanding teaching, both by the American Academy of Environmental Medicine. She was a recipient of the prestigious Hind Ratna award by an NRI organization in India. She is a co-author of the medical textbook, *Reversibility of Chronic Degenerative Disease and Hypersensitivity, V. 1: Regulating Mechanisms of Chemical Sensitivity.* Dr. Patel has published many peer-reviewed research papers related to the topic of environmental medicine. Dr. Patel currently serves on the board and is president of the Environmental Health Foundation of New York. She was a president of the American Board of Environmental Medicine and previously served on the board of the American Academy of Environmental Medicine. She previously held the position of Director of Child Health at Department of Health Erie county and chief of pediatrics at Deaconess Hospital, Buffalo, New York. Dr. Patel is a fellow of the American Academy of Environmental Medicine.

Dr. Patel was born in Poona, India and was raised in Ahmadabad, India. Dr. Patel graduated from St Xavier's college with honors in the state of Gujarat, India and also with honors from B. J. Medical College, Gujarat University in Ahmadabad, India. She then completed a rotating internship at Bexar County Hospital in San Antonio, Texas. She held a pediatric residency from 1969 to 1972. Dr. Patel is an assistant professor of pediatrics at the State University of New York at Buffalo since 1973.

Acknowledgments/ Preface

THIS BOOK IS DEDICATED to Robert Becker, MD, an orthopedic surgeon who was one of the first clinicians to recognize the significance of EMF in medicine and surgery, and to his assistant, Andrew Marino, PhD, who helped work out the basic science of orthopedic electromagnet healing.

Recognition goes to the cardiac physiologists and surgeons who worked out the basic tenets of cardiac physiology, who have the greatest advances in cardiac surgery.

This book is further dedicated to the great physiologists like Hartmut Heine, PhD, who discovered the new field of electromagnetic sensitivity—the new epidemic of the twenty-first century. This epidemic follows the problem of chemical sensitivity, for which there are 80,000 chemicals with combinations too numerous to count.

The combinations of electrical and chemical problems are so multitudinal that they are staggering to the clinician. However, a cursory understanding of the combinations of these fields can help clinicians partially understand the problems and thus help in the diagnosis and treatment of electromagnetic sensitivities.

Eventually, new treatments can be developed, eliminating illness, as shown for the healing of bones. A basic understanding of environmentally induced illness and healing must be reached by clinicians before fixed named diseases occur.

We dedicate this book to Professor Cyril Smith, PhD, at the University of Salford in England as well—along with Dr. Jean Monro, MD, a clinician par excellence, also founder and director of the Breakspear Environmental Hospital in England. She is not only Dr. Rea's colleague and dear friend but is an innovator in keeping environmental medicine alive on the continent of Europe.

Randolph T. and Dickey L. defined the total body pollutant with chemical sensitivity that laid the foundation for understanding increased deficiency and electrical sensitivity for developing their less polluted environmental unit.

Finally, to Sam Milham, who pointed out the effect of dirty electricity on schoolchildren and teachers.

CHAPTER 1

Air Pollution

INTRODUCTION

The entity of electromagnetic sensitivity has been a mystery in modern medicine yet has been known commonly for years. It has been well known that the body runs off electromagnetic forces since the days of the discovery of electricity. People have been electrocuted by dirty electricity and eventually shocked back to life. We now have technology that is mainly dependent upon electricity, such as EKGs to measure heart rhythms, EEGs to measure brain waves, EMGs to measure muscle rhythms, and so on.

It is a known fact that shocking a patient back to life happens when the heart develops a fatal arrhythmia. A lot of physicians don't know about the total body pollutant load, which is the sum total of the pollutants (i.e., pesticides, natural gas, solvents, mycotoxins, bacteria, viruses, and other injuries the patient takes in to function) throughout life. Eighty percent of people who are EMF sensitive have food and chemical sensitivities.

LOCATION ON EARTH

Along with emissions and weather conditions (meteorological phenomena), location on earth is extremely important when evaluating outdoor environment and EMF sensitivity. This evaluation has to be local for where the individual patient lives and works. Most large urban areas in the United States are located near the sea, lake coastal areas, or rivers, and much heavy industry is located in valleys, for example, the Ohio River Valley; Charleston, West Virginia; Pittsburgh, Pennsylvania—Allegheny & Monongahela River Valley, and so on. *The local airflow or lack of it in these regions has a significant impact on the pollution dispersal processes and thus the total body pollutant and EMF load.* As shown later in this section, breeze changes from land to sea and sea to land can help or hinder dispersal of pollutants in valleys. Industry in valleys is a setup for inversions at night because of low horizontal airflow; that is, Mexico City is a high mountainous area set in a low lake bed that holds in the pollution, Pittsburgh, Pennsylvania, is in a low river valley, and so on).

This environmentally compromised condition has been observed to exacerbate both periodic (chemical sensitivity) and aperiodic (chronic degenerative diseases) homeostatic disturbances and prevent people from being able to clear the early onset of maladies. There is clearly a higher incidence of disease in industrialized valleys.[1]

Some locations, more than others, tend to accumulate outdoor air pollutants. Low-lying areas such as valleys, low-rolling plains, or canyons tend to accumulate pollutants because dispersal is difficult and the heavier pollutants settle and are able to concentrate in those areas. Also, areas located in a wind pattern from pollution generators, for example, freeways,[1] factories,[1] and toxic dumps,[1] or aerial insecticide spraying and gas, oil fields, fields with sulfur content, and refineries, are additionally polluted by winds carrying emissions from those sources.[1]

High mold and algae areas are found near large bodies of water and where the air is always damp, like in the Louisiana bayous. They may disrupt the dynamics of homeostasis in many individuals. The red tide is an example of this kind of air pollution problem.[1] At times, in the United States in the gulf coast of Texas and the southeast coast, the red tide (algae) will move in quickly, up to elevations of 1000–2000 feet from the water, killing the sea life and causing bronchial irritation with coughing and neurovascular phenomenon like confusion, short-term memory loss, and loss of balance in humans. This type of exposure has also been known to produce psychosis in some people. Seashores with high pollutant generators tend to clear the air easily, although when weather inversions occur, there may be high levels of pollution in the area for days. Inversions with no wind movement for days can be generators of ill health. *They can at times incapacitate the individual with chemical and electrical sensitivity due to the massive increase in total environmental pollutant load.*

Diurnal Patterns

In one study, Hildebrandt compared the diurnal transport of lower atmospheric pollution in Phoenix, Arizona (USA), with the daily variability of air pollution in another urbanized valley, Kathmandu, Nepal. He found evidence that the physical setting of Kathmandu partially accounts for the diurnal variability of pollutant levels in the capital city of the Kingdom of Nepal.[2]

Past studies have investigated the climatic conditions that promote high pollution levels in Phoenix, Arizona.[3,4] Likewise, scientists have examined the anthropogenic and natural factors that partially account for daily concentration variability and the transport of air pollutants throughout the Phoenix Valley.[4,5]

In several respects, Kathmandu, Nepal, is similar to Phoenix in that both are affected by distinct monsoon seasons with high summer temperatures and moderate winters. Both are densely populated; have a high reliance on personal motor vehicles; and lie within large, urbanized valleys. While Phoenix is nearly surrounded by mountains and the effects of the topography on air pollution transport are well documented,[4–6] there is little such documentation for Kathmandu. In fact, few quantitative studies have been performed in Kathmandu to examine the climatic conditions that promote high levels of air pollution,[3,8] and no studies have been performed to examine the influence of the local topography on the transport of air pollution in and around Kathmandu Valley.

PM_{10} is defined as particulate matter (PM) with a diameter of less than 10 micrometers (μm). PM_{10} is therefore very small, remains suspended in the air for periods of time, and is easily inhaled into the deep lung. Hourly measurements of atmospheric PM_{10} have been available for the Kathmandu area since 1999, providing an opportunity to characterize temporal patterns of air pollution from another urbanized valley and compare the results to temporal variations reported in Phoenix. Identifying climatic factors that control variations in urban pollution levels may further illuminate how the topography affects pollution concentration and transport. Analysis of Kathmandu, particularly when compared to Phoenix, may expand our understanding of the dynamics of pollution levels in urbanized valleys.[2]

METEOROLOGICAL PHENOMENA

Meteorological phenomena are the third part in the evaluation of outdoor air contaminants combined with emissions and location on the earth in relation to the patient with chemical sensitivity or chronic degenerative disease.[9] There are some general observations that one should be looking for when there is a diagnosis of outdoor air contamination. One should remember that the earth is at the mercy of spinning (earth's rotation) and fire (the sun). Because of these two phenomena, patients with chemical and electric sensitivity and chronic degenerative disease will have many varied responses because of the fragility of their receptors, which can be sensitized and damaged by pollutant exposures and/or nutrient deficiencies. This phenomenon can result in electrical sensitivity.

Magnetic Fields

The earth is one of the planets that has a strong magnetic field (Figure 1.1). In the absence of any external forces, the geomagnetic field can be approximated by a dipole field with an axis tilted about 11° from the spin axis.[9]

Magnetic Fields

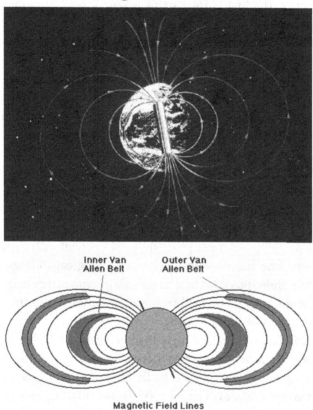

FIGURE 1.1   Magnetic fields of the earth showing (with) the dipole field
with the axis tilted 11° from the spin axis. (From NASA.)

The force created by the solar wind is able to modify this magnetic
field, creating a cavity, the magnetosphere. The magnetosphere is
the region of space to which the earth's magnetic field is confined
by the solar wind plasma, blowing outward from the sun, extending
to distances of 60,000 kilometers from earth. This field has a
profound influence on life on earth and the patient with chemical
and electrical sensitivity and/or chronic degenerative disease.

The earth's magnetosphere is formed from two essential ingredients. The first is the earth's magnetic field, generated by currents flowing in the earth's core due to its iron content. Outside the earth, this field has the same form as that of a bar magnet, a dipole field aligned approximately with the earth's spin axis. The second ingredient of the earth's magnetosphere is the solar wind, a fully ionized hydrogen/helium plasma that streams continuously outward from the sun into the solar system at speeds of about 300–800 kilometers per second. *The clinician must consider the effects of the earth's magnetic field when studying an individual with dyshomeostasis and deranged metabolism because many of these individuals have derangements related to EMF changes.*

The wind is composed of protons and alpha particles, together with sufficient electrons that are electrically neutral overall. Any of these particles can create problems for the patient with chemical sensitivity who has damaged electrical coping capabilities. The solar wind is also preceded by a large-scale interplanetary magnetic field, the solar magnetic field transported outward into the solar system by the solar wind plasma.

There is, however, a third ingredient in the essential elements of the earth's magnetosphere. This is the earth's ionosphere. The upper atmosphere is partially ionized by far ultraviolet and x-rays from the sun above altitudes of about 100 km. The resulting ionosphere consists of mainly protons, singly charged helium and oxygen, and the requisite number of electrons for electric charge neutrality.

The solar wind plasma is frozen to the interplanetary magnetic fields and the earth's plasma (for example, from the ionosphere) to the earth's field. When these plasmas meet each other, they do not mix but instead form distinct regions separated by a thin boundary. The solar wind thus contains the earth's magnetic field in a cavity surrounding the planet.

The size of the cavity of the magnetosphere is then determined by pressure balance at the boundary between the solar wind on one side and the magnetic pressure of the planetary field on the other. The planetary and "bar magnet" field produces a field strength of about 30,000 nT of the earth's surface at the equator. This physical phenomenon may well influence health in a compromised individual such as the patient with chemical and electrical sensitivity and/ or chronic degenerative disease. Estimates place the boundary at the geocentric distance of about 10 earth radii (64,000 km) on the upstream side, and on the downstream side, the cavity extends into a long magnetic foil. Across the magnetopause, the magnetic field usually undergoes a sharp change in strength and direction. A sheet of electrical current must flow in the plasma in this interphase. A shock wave also stands in the flow upstream of the cavity, which forms because the speed of the solar wind relative to the earth is much faster than the wave of propagation within it. Across the shock, the flow is slowed, compressed, and heated, forming a layer of turbulent plasma (magnetosphere) outside the magnetopause. Inside the cavity, the plasma sphere will approximately rotate with the earth. This will occupy some space because the earth's fields are frozen into the ionosphereic plasma. The corotation is enforced in the absence of other driving forces by collisions between ions and atmospheric neutrals at heights of about 120–140 kilometers.

The aurora, or northern and southern lights, are often visible from the surface of the earth at high northern and southern latitudes. Auroras typically appear as luminous bands or streamers that can extend to altitudes of 200 miles, well into the ionosphere. Auroras are caused by high-energy particles from the solar wind that is trapped in the earth's magnetic field. The collision of trapped particles with atmospheric molecules causes spectacular effects in the visible spectrum, but these excited molecules can also emit radiation in other wavelength bands like ultraviolet, x-ray, and so on. If any of these electric and magnetic phenomena get into the earth's atmosphere, they will affect the chemically and electrically sensitive and chronic degenerative diseased patient.[9]

Solar Wind

The solar wind streams off the sun in all directions at speeds of about 400 km/s (about 1 million miles per hour). The source of the solar wind is the sun's hot corona. The temperature of the corona is so high that the sun's gravity cannot hold onto it.

The solar wind is not uniform. Although it is always directed away from the sun, it changes speed and carries with it magnetic clouds, interacting regions where high-speed wind catches up with slow-speed wind, and composition variations. The solar wind speed is high (800 km/s) over coronal holes and low (300 km/s) over streamers. These high- and low-speed streams interact with each other and alternately pass by the earth as the sun rotates. These wind speed variations buffet the earth's magnetic field and can produce storms in the earth's magnetosphere. Once this occurs, effects may be seen in chemically sensitive and chronic degenerative disease patients. Patients may become depressed or develop unstable physiology, worked on in many cases of magnetic wares, which can result in electrical sensitivity.

A map of some of the major regions in our earth's magnetosphere has been identified.[10] When the solar wind comes to the earth, it meets the earth's magnetic field. Most of the tiny particles in the solar wind are pushed around the earth because of this magnetic field. They begin their journey around in a curve called the "bow shock." Just like water makes a curved wave in front of a boat, the solar wind makes a curve in front of the earth. After passing through a shock wave at the bow shock, the wind flows around the magnetosphere and stretches it into a long tail. However, some solar wind particles leak through the magnetic barrier and are trapped inside. Solar wind particles also rush through funnel-like openings (cusps) at the North and South Poles, releasing tremendous energy when they hit the upper atmosphere. The northern and southern lights (auroras) are the evidence we can see of this energy transfer from the sun to the earth. The particles then follow a path that goes around the earth in a sort of cover or sheath. This curve is the "magnetosheath." These particles mix with other particles that come up from the earth's ionosphere to fill the magnetosphere (see Figure 1.2, a map of the magnetosphere, with sun-earth light transfers).

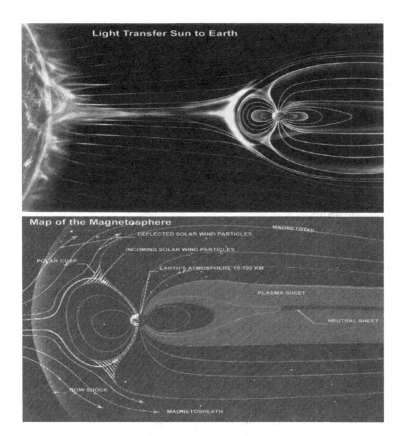

FIGURE 1.2   Transfer of light from sun to earth. Map of magnetosphere. (From NASA.)

A great many of the tiny pieces of matter in the earth's magnetosphere do not come from the sun's solar wind. They come from the earth. Our planet can lift air particles into space, and these particles will become charged by the time they get there. *The lifting of particles creates movement that, if inhaled, could possibly cause damage to human health.* Certainly, the environmentally damaged patient will perceive and react to these atmospheric alterations. We actually have developed an antigen extract from the sun's rays that seems to help some individuals who are sun sensitive.

The solar wind travels past the earth at well over 1,000,000 miles per hour, and, thanks to the earth's magnetic field, the solar wind is stopped and deflected around the earth so that most of it does not hit our atmosphere head on.[10]

Ultraviolet rays from the sun ionize the upper atmosphere, creating the electrically conducting ionosphere and a source of plasma for the magnetosphere. The energy from the solar wind enters the magnetosphere and is stored there, being released in sudden surges, which can make the electrically sensitive worse.

The Sunspot Cycle

Sunspots appear as dark spots on the surface of the sun. Temperatures in the dark centers of sunspots drop to about 3700 K (compared to 5700 K for the surrounding photosphere). They typically last for several days, although very large ones may live for several weeks. Sunspots are magnetic regions on the sun with magnetic field strengths thousands of times stronger than the earth's magnetic field. Sunspots usually come in groups, with two sets of spots. One set will have a positive or north magnetic field, while the other set will have a negative or south magnetic field. The field is strongest in the darker parts of the sunspots, the umbra. The field is weaker and more horizontal in the lighter part, the penumbra.

The sunspot number is calculated by first counting the number of sunspot groups and then the number of individual sunspots. The "sunspot number" is then given by the sum of the number of individual sunspots and 10 times the number of groups. Since most sunspot groups have, on average, about 10 spots, this formula for counting sunspots gives reliable numbers even when the observing conditions are less than ideal and small spots are hard to see. Averages (updated monthly) of sunspot numbers show that the number of sunspots visible on the sun waxes and wanes with an approximate 11-year cycle.

Early records of sunspots indicate that the sun went through a period of inactivity in the late seventeenth century. Very few sunspots were seen on the sun from about 1645 to 1715. This period of solar inactivity also corresponds to a climatic period called the "Little Ice Age," when rivers that were normally ice-free froze and snow fields remained year-round at lower altitudes. There is evidence that the sun has had similar periods of inactivity in the more distant past. The connection between solar activity and terrestrial climate is an area of ongoing research.

These data show that sunspots do not appear at random over the surface of the sun but are concentrated in two latitude bands on either side of the equator. A butterfly diagram showing the positions of the spots for each rotation of the sun since May 1874 shows that these bands first form at midlatitudes, widen, and then move toward the equator as each cycle progresses. The cycles overlap at the time of sunspot cycle minimum, with old cycle spots near the equator and new cycle spots at high latitudes. Studies have shown that human behavior, at times, correlates with levels of sunspots. The electrically sensitive patient may be influenced by these cycles.

There is an 11-year cycle of a close relationship between the number of sunspots and the potential gradients; the latter increases 20% with a 100% increase in the number of sunspots. The daily and yearly amplitudes of the potential gradient also fluctuate parallel to the curve of sun activity and seem to be related to the northern lights phenomenon. *There has been an increase in the effects of sunspots over the last two to several years. This increase, along with an increase in human-made electrical radiation, may be the reason for the increase in the number of severely affected electrically sensitive and chronic degenerative disease individuals.*

Solar Flares

Solar flares are tremendous explosions on the surface of the sun. In a matter of just a few minutes, they heat material to many millions of degrees and release as much energy as a billion megatons of TNT. They occur near sunspots, usually along the dividing (neutral) line between areas of oppositely directed magnetic fields.

Flares release energy in many forms—electromagnetic (gamma rays and x-rays), energetic particles (protons and electrons), and mass flows. The images in are from the Big Bear Solar Observatory. The upper image shows material erupting from a flare near the limb of the sun on October 10, 1971. The lower image shows a powerful flare observed on the disk of the sun on August 7, 1972. This is an example of a "two-ribbon" flare in which the flaring region appears as two bright lines threading through the area between sunspots within a sunspot group. This particular flare, the "seahorse flare," produced radiation levels that would have been harmful to astronauts if a moon mission had been in progress at the time.

The key to understanding and predicting solar flares is the structure of the magnetic field around sunspots. If this structure becomes twisted and sheared, then magnetic field lines can cross and reconnect with the explosive release of energy. , The dark lines represent the neutral lines between areas of oppositely directed magnetic fields. Normally, the magnetic field would loop directly across these lines from positive (outward-pointing magnetic field) to negative (inward-pointing magnetic field) regions. The small line segments show the strength and direction of the magnetic field measured with the Marshall Space Flight Center (MSFC) vector magnetograph. These lines and line segments overlie an image of a group of sunspots with a flaring region. Again, these flares, which influence earth's lower magnetosphere, could affect health.

The most powerful solar flare of the year erupted from the sun April 11, 2013, sparking a temporary radio blackout on earth, NASA officials say.

The solar flare occurred at 3:16 a.m. EDT (0716 GMT) and registered as a M6.5 class sun storm, a relatively midlevel flare on the scale of solar tempests. It coincided with an eruption of super-hot solar plasma known as a coronal mass ejection.

Increased numbers of flares are quite common since the sun's normal 11-year cycle is ramping up toward solar maximum, which is expected in late 2013.

The April 11 M-class solar flare was about 10 times weaker than X-class flares, which are the strongest flares the sun can unleash. M-class solar flares are the weakest solar events that can still trigger space weather effects near earth, such as communications interruptions or spectacular northern lights displays.

The solar flare triggered a short-lived radio communications blackout on earth that registered as an R_2 event (on a scale of R_1 to R_5), according to space weather scales maintained by NOAA.

When aimed directly at earth, major solar flares and coronal mass ejections can pose a threat to astronauts and satellites in orbit. They can interfere with GPS navigation and communications satellite signals in space, as well as impairing power system infrastructure on earth.

Substorms

The earth's magnetosphere, like the earth's atmosphere, is never at rest. Of its many dynamic features, perhaps the most important and basic is the *magnetospheric substorm*, a period of the order of 1 hour or less, during which energy is rapidly released in the magnetospheric tail. During substorms, in the polar regions, aurora becomes widespread and intense, as well as much more agitated, and the earth's magnetic field is disturbed. At times, patients with damaged electromagnetic receptors tend to worsen to the point of incapacity.

Magnetic storms decrease in the magnetic field worldwide and are readily observed at any place. Typically, a storm takes about half a day to develop, and it gradually decays over the next few days. *This phenomenon may be one of the reasons the patient with chemical and electromagnetic sensitivity may have intermittent episodes of instability in which the sensitivities are exacerbated and the intradermal endpoints altered.*

Magnetic storms are relatively rare. On the other hand, smaller "substorms," observable mainly in polar regions (and in space), present a clearer pattern and seem to be more fundamental. They are also much more frequent, often just hours apart.

Storms distinguish themselves by injecting appreciable numbers of ions and electrons from the tail into the outer radiation belt, and their worldwide magnetic disturbance reflects a rapid growth of the ring current. Substorms usually do not inject as many particles as intense storms. It might thus be that magnetic storms are merely sequences of very intense substorms, but additional factors are also involved; in particular, magnetic storms require external stimuli such as the arrival of a shock front or a fast stream in the solar wind.

On earth, the most visible sign of a substorm is a great increase of polar auroras in the midnight aurora zone. At ordinary times, quiescent aurora arcs are often seen there, but following the onset of a substorm, they intensify, move rapidly (mostly poleward), and expand, until they may cover much of the sky. Their activity may build up for half an hour and then decay, but, as with atmospheric weather, patterns are quite variable.

Large magnetic disturbances are also observed, up to 1000 nT (nanotesla), which is about 2% of the total field in the aurora zone. The worldwide disturbance observed in a magnetic storm of respectable size may only reach 100 nT, but then, its source is much more distant, namely, the ring current, which circles the earth at distances of tens of thousands of kilometers. The electric currents associated with the substorm, on the other hand, come down to the ionosphere, only about 130 km above the ground .

Most natural phenomena require an input of energy, which is then changed to some other form. This also holds true for substorms. It seems no accident that they generally occur when the interplanetary magnetic field (IMF) has a southward slant, which, as noted in the discussion of the open magnetosphere, is a time when interplanetary field lines might be more strongly linked to those of the earth and more energy flows from the solar wind to the magnetosphere.

This is also a time of faster "reconnection" between interplanetary and terrestrial field lines, a time of more rapid "peeling away" of magnetic field lines from the day side (together with their attached plasma), as they become attached to interplanetary field lines and are dragged with them into the tail. Any overall plot of magnetospheric field lines shows the lines parted like combed hair; one group closes on the day side, around noon, another group is pulled back into the tail lobes, and the cusp marks the groups' separation. Increased peeling away near noon shifts the balance: fewer lines go sunward, more into the tail, and the cusp shifts to a field line anchored closer to the earth's equator. Such a shift leads to two effects.

First, it weakens the earth's magnetic field near noon, where plasma and field lines have been peeled off, allowing the solar wind to push its way closer to earth. As a result, when the interplanetary magnetic field is southward, the "nose" of the magnetosphere is seen to be (on the average) about 1 RE further earthward than with northward IMF, out of a mean distance of 10–11 RE.

Second, more of the magnetic field is drawn into the tail, and the tail lobes expand, storing additional magnetic energy in them. It is widely believed that the expanded lobes are the main storehouse of energy, which powers the substorm. Sometimes, in "clean" substorms when the IMF suddenly "turns southward" after a long quiet period, one can observe this reservoir of energy charging up, as the tail field intensifies and magnetic field lines in synchronous orbit become increasingly stretched tailward, like a slingshot. This "growth phase" typically lasts 40 minutes. The exact way in which this energy is released and the "trigger," which starts the process, are still subjects of debate and controversy.

General Finding

A vast ocean of air, the atmosphere, envelops the earth. The atmosphere's depth is 150 km, with half of its weight being in the lowest 1 km. The air is denser in the winter than in the summer, which tends to add to air pollution. Patients with chemical and electrical sensitivity have more problems with the dense air. The lower part of the atmosphere is in constant turmoil (generating electric and electromagnetic fields of various types, as shown in the previous section) and denser than the upper atmosphere, as shown in the previous pages. Radiation from the sun consists of electromagnetic waves that take many forms, traveling at the speed of light, 300,000 km/sec. Light takes 8 minutes to reach the earth from the sun. The sun radiates energy at a constant rate (solar constant) of 1.35 kW/m^2 that gives the clinician a tool to evaluate some types of pollutant dispersion and its electromagnetics along with other effects upon humans. Clouds, dust, smoke, buildings, and trees deplete solar energy by absorption and distortion.

By trapping some of the solar outgoing radiation from the earth, the atmosphere keeps the earth's surface warmer than it otherwise would be. This trapping has increased markedly over the last few years due to excess pollution that causes the formation of more intense clouds (greenhouse effect) and thus starts a vicious cycle of more pollution.

Concrete cities create islands of their own weather by absorbing heat all day and emitting it at night. This phenomenon often influences the cities' own weather and pollutant cycles by causing overheating and local wind and inversion patterns, the results of which can be seen in early morning haze and pollution caps that are visible when one drives into or flies over a major city in the early morning. These local inversion and wind patterns create pollution caps over the city, worsening their own air pollution problems and those of places downwind from them (Figure 1.3). Obviously, the local weather patterns can have a dramatic effect on the chemically sensitive or chronic degenerative disease patient's well-being and thus evaluation of these factors must be done.[1]

In the summer, when the sun's rays strike the earth at a more direct angle, the temperature becomes hotter. Higher temperatures occur on longer days with more vertical rays from the sun. This process will tend to evaporate more pollution, but it also creates more ozone as the sun's rays hit auto exhaust. *Clearly, ozone frequently exacerbates the problems of the chemically sensitive and chronic degenerative disease patient, with some of the symptoms triggered by pollutants generated during some part of the day.*

Concrete Cities Create Pollution Caps Caused by Local Inversion and Wind Patterns

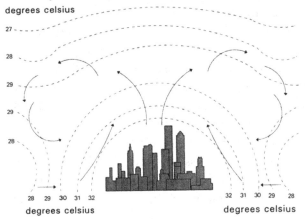

FIGURE 1.3   Country breeze flowing at low levels toward an urban heat island. Temperatures along the isotherms are indicated in degrees Celsius.

There is enough pollutant burn-off at times, especially when the heat is hot enough on the earth, to relieve chemical sensitivity or chronic degenerative disease. Chemically sensitive and chronic degenerative disease patients generally do much better in the summer, particularly if they can remove themselves from the main pollutant generators. However, other chemical sensitivities in different stages do worse with the heat.

Air does not flow smoothly over the earth's surface, thus creating many changes, which will influence the diagnostic response of the chemically and electrically sensitive and chronic degenerative disease patient. Rather, it follows a three-dimensional movement of "turbulence," which not only results in air movement change but also in changes in the quantity or quality of air ions (particularly positive ions) and other electric and electromagnetic phenomena. This movement is produced by thermal and mechanical turbulence. Thermal turbulence results from the atmosphere heating, while mechanical turbulence results from the movement of air past an obstruction. Thermal turbulence is dominant on clear sunny days with light winds. The effect of this turbulence is to enhance the dispersion process, which helps the chemically and electrically sensitive and chronic degenerative disease patient. In the case of mechanical turbulence, downwash from the pollution source may result in high pollution levels downstream. If the chemically sensitive or chronic degenerative disease patient lives and/or works in these types of high-pollution and turbulence areas, he/she often has many functional difficulties like aches, pains, mood swings, and energy bursts, followed by fatigue.

Different surfaces absorb heat at different rates. Snow absorbs heat by 25% of the total rate possible, water by 60%–96%, dry sand by 75%, and plowed fields by 80%–90%. The lighter the color of the surface, the less able it is to absorb, but other factors are also involved. Oceans, for example, cool and warm more slowly than land because of the differences in the properties of liquids and solids. Patients with chemical and/or electrical sensitivity and/or chronic degenerative disease generally do better near large bodies of water if they are not too sensitive to mold or algae and if there are low pollution emissions. When this type of living area can be obtained, the chronically ill patient becomes more stable and the diagnosis of the environmental part of the illness will be easier to manage and treat.

Weather cycles depend on bursts of fire from the sun (sunspots), which create ion streams in space, which, in turn, help create pressure changes and ion changes on the earth. When the air mass near the earth's surface becomes slightly heated, it will rise because it is lighter than the surrounding air (Figure 1.4).[1]

Whether the air mass will go on rising depends on the temperature of the surrounding air, which changes with height. If the heat of the air near the surface of the earth is maintained, there will be a continuing upward current, the thermal. Thermal areas are pockets of better air that tend to sweep pollutants up and out. One simple indicator for finding a thermal area is to look for soaring birds that spread their wings motionless and spiral upward. When the patient with chemical or electrical sensitivity and/or chronic degenerative disease finds these areas, he/she tends to function

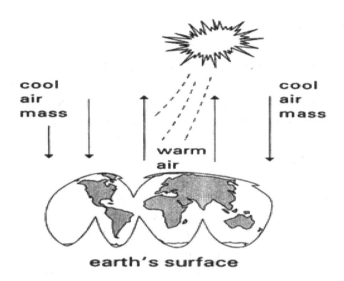

FIGURE 1.4   Thermals—pockets of less-polluted air. (Rea, W. J. 1994. *Chemical Sensitivity. Sources of Total Body Load.* Vol. 2. Boca Raton, FL: Lewis, pp. 655.[1]) (May be a small area of 1000 ft or a large area of up to 5 mi.)

better. For example, we are familiar with one environmentally controlled colony that is built in one of these thermal areas where some of the sickest pollutant-wounded patients rapidly improve.

Moisture condensing out from rising cool, wet air becomes a visible cumulus cloud that marks the top of a thermal. If the upward current goes on rising to very great heights, the result is a cumulus thundercloud. Ground air pollution can be swept up and out by this phenomenon, or it can accumulate as a thundercloud and dump pollutants on places farther away. This phenomenon often results in acid rain for areas further to the east; if pollutants are high at the source, this thermal cloud phenomenon occurs. Such emitters as power plants, highways, airports, factories, insecticide-sprayed fields, and cities generate this excessive pollution. *The patient with chemical or electrical sensitivity and/or chronic degenerative disease will have a more labile metabolism in these contaminated areas, making the dynamics of environmental overload more difficult to perceive and obtain for diagnostic purposes.*

Atmospheric Pressure

The various effects of atmospheric pressure and winds on climate are mostly due to the cold atmosphere (cold jet stream) from the poles to the equator and the hot atmosphere spreading from the equator to the poles (subtropical jet stream).[1] (See Hadley's model—Figure 1.5.) Pollutant emitters should always be noted in order to evaluate the amount and types of pollutants carried by these winds.[1]

At the equator, intense heating of land and water occurs, warming the air above. This warm air rises to a high elevation, proceeding toward the pole. As the air mass reaches the pole, it loses its heat by radiation and cooling and convection, which subside and turn toward the equator to complete the cycle. Hadley's model does not describe actual air circulation, since other factors such as the Coriolis effect of the rotation of the earth and friction influence global air movement. Here, the eastward rotation of

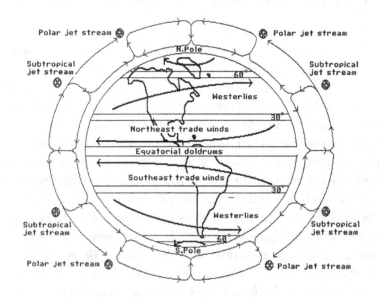

FIGURE 1.5 Schematic representation of general circulation of air currents able to affect the chemically sensitive. Hadley's model.

the earth deflects the winds originating from the north at their meeting point, toward the west.[1]

The Coriolis effect is that large-scale air movement north or south that appears to be deflected from its expected course. For example, in the northern hemisphere, air moving toward the North Pole appears to move east. Air moving to the south appears to move west. A converse relationship exists in the southern hemisphere. This apparent deflection of large-scale movements is due to the earth's rotation.

Most of the winds in the continental United States come from the west and blow east, except in the southern United States, where their origination constantly switches from the southeast (due to trade winds from the subtropical high) to the west. The northern direction of the winds in these areas depends on the time of year with the tilt of the earth. In addition, polar (formed at the convergence of polar and temperate air masses) and subtropical (formed at 30° latitude, where the tropical and temperate air masses converge) jet streams blow out of the north and from the equator, respectively, often compounding, but sometimes relieving, the pollution problem on earth. Again, these cleaning or polluting winds depend on the quality of emitters in the area or region. Steady, nongusting winds over 5 mi/hr seem to be best at dispersing pollutants.

An artificial electromagnetic jet stream exists across the eastern United States, apparently occurring from artificial sources. Cell towers are being placed all over the United States in order to service cell phones, and these are making electromagnetic pollution worse. Adverse effects of this phenomenon are only now becoming apparent as a potential health problem. *Many more people are reacting to electromagnetic radiation.* Magnetic lines, being wider over the equator, stretch from the poles. These areas of magnetic radiation may also adversely influence local areas of air pollution and thus have a negative impact on electrically and chemically sensitive individuals as well as those with chronic degenerative disease. Diagnosis of these contaminants is done through meters, visual inspection, and particulate counters.

Horizontal movements result from spatial pressure differences, which are produced by unequal heating and cooling of the earth's surface. The pressure of the air was once expressed in millimeters of mercury (690–795 being normal). Now it is in millibars (mb); therefore, 0.00 mb is the average pressure at sea level. Points of similar pressure connected together on a map are isobars. The patterns we will discuss include depressions and fronts, cyclones and fronts, and inversions and how they apply to the evaluation of the noxious incitant environmental load.

Electric Properties of the Atmosphere

Many people are weather sensitive (Table 1.1). Estimates are as high as 30%–40% of the population.[1] However, the whole population has to deal with the weather and weather changes that affect almost all patients with chemical and/or electrical sensitivity and chronic degenerative disease. Since many patients with chemical sensitivity and chronic degenerative disease are electrically sensitive and most seem to be influenced by some form of electricity, a small section on atmospheric electrical phenomena is in order. Ion streams come from the sun hitting the earth or are generated by wind shearing off mountains, deserts, seashores, or waterfalls. The earth's own electromagnetic properties combined with the generation of artificial jet streams necessitates that severe electrical phenomena be considered and related to the malfunctioning of chemically sensitive and chronic degenerative disease patients.

The electric and magnetic properties of the atmosphere manifest as ionization and electric conductivity and electromagnetic long waves (Sferics; 6–100 km) at electric fields.

Ionization and Electric Conductivity

Air ion formation begins when high atmospheric energy acts on a gaseous molecule to eject an electron. The displaced electron attaches itself to an adjacent molecule, transforming it into a negative ion. The original molecule, deprived of its balanced dipole, becomes a positive ion. The atmosphere contains positive and

TABLE 1.1   Weather Complaints Reported by 778 Subjects in Switzerland
(V. Faust)

Complaints	All Subjects (%)	Weather-Sensitive Subjects (%)	Non–Weather Sensitive Subjects (%)
Impaired concentration	23	37	8
Forgetfulness	15	24	6
Confusion	15	23	7
Fatigue	39	57	21
Exhaustion	12	22	3
Ill-humor	31	48	14
Nervousness	29	30	9
Anxiety	8	15	2
Depression	10	18	3
Headache	28	44	13
Migraine	23	38	7
Ill-feeling	12	22	3
Vomiting	2	3	0
Sleeplessness (insomnia)	27	42	14
Initial insomnia	23	35	12
Terminal insomnia	16	23	9
Sweating palms	10	15	5
Hot flushes	8	13	4
Convulsive tics	5	9	1
Chills	2	5	1
Dislike of work	3	45	16
Inability to work	3	5	1
Lack of appetite	9	15	3
Diarrhea	2	4	1
Frequent micturition	4	7	2
Hearing disturbances	2	5	0
Smell disturbances	4	6	1
Taste disturbances	4	6	1
Visual disturbances	15	24	6
Skin sensitivity	5	8	1
Vertigo	13	23	4
Faintness	2	4	1
Heart palpitations	12	20	4

(Continued)

TABLE 1.1 (Continued)   Weather Complaints Reported by 778 Subjects in
Switzerland (V. Faust)

Complaints	All Subjects (%)	Weather-Sensitive Subjects (%)	Non–Weather Sensitive Subjects (%)
Breathlessness	10	16	3
Scar pain	12	19	5
Rheumatoid pain	9	15	3
Bone fracture pain	17	26	8

Source: Sulman, F. G. 1976. *Health, Weather and Climate* P. 89, Basel, Switzerland:
    S. Karger. With permission.[140]

negative ions, but the potential pollutant gradient falls off as the
atmosphere is ascended, suggesting that there must be an excess
of positive ions. Indeed, the tendency toward more positive ions
enhances the effects of air pollution. Most particulate air pollution
is car exhaust and coal dust, which are predominately positive ions.

The atmosphere also contains a large number of electrically
charged oxygen and water particles that are often classed into
three groups: small, intermediate, and large ions. Small ions
are composed of electrically charged clusters of molecules of
atmospheric gases (0.001–0.003 μm) with great mobility (km/sec)
and an electric charge per ion of 15.9×10^{-20} coulomb. There is
an average of 500 negative and 600 positive small ions in 1 cm³ of
air that contains a vast number of ordinary molecules (2.75×10^{19}
in 1 cm³). They decay within a few seconds. Intermediate ions
contain large clusters of 0.003- to 0.03-μm diameters, with a slower
mobility (0.5 cm/sec). Large ions are sluggish ions attached to dust
particles or hygroscopic particles on which moisture condenses
when clouds are formed (average velocity 0.005–0.0005 cm/sec).
In places where the atmosphere is polluted by dust and products
of coal combustion, large ions are more frequent than small ions.
Values of 50,000 and larger ions are seen near towns.

*The chemically sensitive and people with chronic degenerative
disease have more trouble in areas with excess large ions. Specifically,
they have more fatigue and pain.*

Electric Conductivity

As a result of the presence of ions, the air has a certain electrical conductivity. The different properties or sizes of the ions, the differences in speed, and the presence of conductors usually explain variations in conductivity. If large ions are formed, that is, in a strongly polluted atmosphere at the expense of small ions, the effective conductivity is reduced because of the slower mobility of the large ions, and the air-earth current is maintained by a higher potential gradient, which may adversely affect the chemically sensitive or chronic degenerative disease patient.

Negative vs Positive Ionization

The general mobility of negative ions is greater than that of positive ions. A 1:100 ratio of negative to positive ions has been found in certain gases; the ratio between both velocities of negative and positive ions decreases in humid air (in dry air, about 1:1.54; in humid air, 1:1.09). The ratio of positive to negative ions increases with atmospheric pressure and also with rising temperature, making weather-sensitive people vulnerable.

Negative ions are derived from ionizing sources (e.g., at the seashore, mountaintops, waterfalls, coronal discharges of needles, ozone, etc.). Negative ions are 2000–4000 per cubic meter of air in the countryside.[1] (See Table 1.2.)

Positive ions are created by air brushing on metal or rough surfaces (e.g., sand with heat); burning candles, matches, iron stoves, and heating pipes; indoor air-handling systems; radioactive emanations from the surface of the earth; cosmic ultraviolet radiation in the atmosphere; and friction electricity (electricity between ice crystals in snow). *Individuals with chemical and/or electrical sensitivity or chronic degenerative disease generally feel better with the presence of excess negative ions.* They feel well at waterfalls, seashores, and mountaintops, provided other emitters are not in excess. However, when exposed to a negative-ion generator, different levels of ion output may affect them adversely. Therefore, care should be used when evaluating the individual

TABLE 1.2   Negative Ions

Location	Per cc of Air
Waterfalls[a]	4000
Mountaintops[a]	4000
Seashore[a]	4000
Clean country side	2000
Busy country street	1000
Busy city street	500
Factories	200
Smog	50

Source: Modified from Sulman, F. G. 1976. *Health, Weather and Climate* P. 89, Basel, Switzerland: S. Karger. With permission.[140]

[a] Areas where the chemically sensitive patient does well.

effects of these ionizers. Some people experience adverse reactions from their forced-air heating and cooling systems due to a positive-ion effect similar to those reactions on high-positive-ion days.

During land rain, 75% of the ions are positive; the electric potential is negative, but most individuals with chemical and/or electrical sensitivity and chronic degenerative disease improve as soon as it begins to rain.

During thunderstorms, the electrical charge is considerably larger than in the case of nonthunderstorm land rain. The potential gradients fluctuate between high positive and negative values, with the latter dominating. *Some more severely afflicted patients who are both chemically and electrically sensitive have been observed to twitch and even have seizures during thunderstorms.*

On meteorologically undisturbed days, electrical conductivity reaches a maximum value during the night, whereas ionization shows the opposite pattern. This pattern seems to be mainly due to the diurnal cycle of heat and turbulence in the atmosphere, which changes the radioactive content of the air near the surface and transports ionized air from the ground to living quarters at midday. *Again, this increase in electrical conductivity may compound nightly radiation inversions, thus making the chemically and/or electrically sensitive and chronic degenerative disease patient worse.*

Apart from daily variations, air conductivity may change due to weather changes or changes in the amount of contamination of the air (e.g., freeways, industrial plants, etc.). Conductivity also changes with the distance from the earth. Its values increase with movement away from the earth's surface.

Electric Fields

The electric field (or the electric potential gradient) is measured in V/m units and changes in the atmosphere from higher levels to ground levels. In good weather, the electric field at the ground is around 120 V/m and increases with the distance from the earth. At about a 12-km height, it only has a few V/m again. The reasons for fluctuations of the gradient are change with the charge of the earth, air-earth current, variable weather conditions, daily variations of the magnetic field, yearly fluctuations, and sunspot fluctuations.

The negatively charged surface of the earth and positively charged atmosphere create a current. Rain, snow, lightning, and sand can bring negative charges to the ground. Of course, the potential gradients depend to a great extent on the meteorological conditions of the atmosphere.

During quiet weather, the electrical gradient is low; it is higher when there is a haze, and very high when certain types of fog prevail. The higher gradients may enhance pollutant effects in the chemically sensitive and chronic degenerative disease individual. The lower gradients in themselves may be another reason chemically sensitive and chronic degenerative disease individuals do better in less turbulence. During land rain, the gradient is negative, but during thunderstorms and snow, the gradient fluctuates, yielding high positive and negative values. The latter gradient yields up to 1000–2000 tons per cubic centimeter of air. Most of these ions are negative, however, and yield electric fields of 10,000 V/m. The potential gradient above the sea is usually low, between 115 and 140 V/m, and near cities and industrial centers, gradients will be high. These higher gradients seem to adversely affect some chemically and/or electrically sensitive and chronic degenerative disease individuals.

Thunderstorm Electricity

Clouds generally contain a very large number of ions (see Figure 1.6). They carry large electric charges with separate electric poles and, accordingly, generate large electric fields. During most of their lifetime, most thunderstorms have a positive-charge center above and a negative-charge center below, and sometimes there is another positive-charge center near their base. Electrification of the clouds occurs very fast in periods of tens of minutes over spaces about 1 km³. The total electric energy of a thunderstorm is about 10^{12}–10^{13} J, the water content 10^{8}–10^{9} kg, and the electric potential versus the ground potential at the negative- and positive-charge

Effects of Thunderclouds on Earth's Electrical Changes

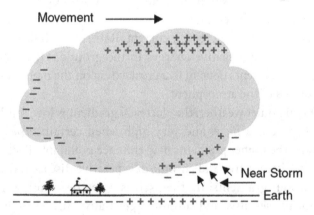

FIGURE 1.6 Effects of thunderclouds on electrical changes on earth often adversely affect the chemically sensitive due to the positive ions of the cloud activity and the negative ions from the earth prestorm. Lightning is also a problem for EMF-sensitive patients. Poststorm, the air is clear with lots of negative ions, and patients feel better.

centers is about plus or minus 10^6–10^7 V. These energy changes will often exacerbate problems in patients who are sensitive to electrical phenomenon and chemicals.

The electric current below the cloud is carried by ions, winds and precipitation, and lightning. The electric field strength at ground level during the storm varies and can reach 10–10^3 V/m or more. *The electrically influenced type of chemically sensitive individual has severe problems during times of intense thunderstorms. Tetany and muscle cramps seem to be high on their symptom list during this period.*

Lightning

Lightning has many different forms and a lifetime of up to a second. Lightning is classified as (1) intercloud, intracloud, cloud-to-ground discharges, and combinations of these; (2) short high-current and long low-current flashes; (3) lightning from the cloud to the ground moving downward, or lightning moving upward; (4) lightning lowering positive or negative charges, or doing both, one after the other.

Typical potential gradients generated by lightning in ground conductors are 10,000 V/m and are quite dangerous. The energy delivered by a lightning strike is about 100 kJ/m.

Long-lasting, low-current lightning is more dangerous to man and more damaging to objects than short high-current lightning. *The chemically sensitive and chronic degenerative disease patient who is electrically sensitive has increased difficulty during episodes of electrical storms.* We have seen patients become severely depressed, demonstrating uncontrolled irritability. They also become very fatigued and weakened, with aching muscles. We have seen several patients develop seizures during lightning strikes who otherwise do well in calm weather.

Sferics (Extra-Long Waves between 6 and 100 km)

Lightning generates a sudden variation of the atmospheric electric field and a fast-rising magnetic field. The corresponding electromagnetic radiation has frequencies from a few hertz through very low frequencies (VLFs) and into very high frequencies (VHFs). The main frequency is about 200 Hz, corresponding to VLF long-wave electromagnetic radiation.

Electromagnetic Activity

There is growing evidence that some human maladies have been due to the discovery and implementation of electricity upon the earth. *Evidence shows that electroduction of the earth is correlated with the evaluation of disease.* This also can be correlated by Lal's opinion that extremely low frequency (ELF) and radio frequency (RF) electromagnetic fields have biological effects (Figure 1.2), which not only can be positive but also can be mildly disruptive to devastating to the human physiology. One can also see that the electromagnetic spectrum stretches from 10–1 to 10–20.

Electromagnetism

Electromagnetism is the increasing prominence of the disease process: a separate section is dedicated to the effects of electromagnetisms. It appears that electromagnetic fields are a prominent factor in generating outdoor and indoor air pollution-triggered malfunctions and disease. It is now clear that dirty electricity from a normal 50/60 volts can cause disease, both hypersensitivity and chronic degenerative types.

Electromagnetic Field Spectrum

The electromagnetic field consists of an electric part and a magnetic part. The electrical part is produced by a voltage gradient and is measured in volts/meter. The magnetic part is generated by any flow current and is measured by Tesla. Magnetic fields as low as .2 ut (a millionth of a Tesla) can produce biological effects.

The electromagnetic spectrum is the range of all possible frequencies of EMF radiation.[11] The EMF spectrum of an object is the characteristic distribution of EMF radiation emitted or absorbed by the particular object. The EM spectrum extends from low frequencies (radio waves) to gamma radiation short wavelengths (high frequencies). The limit of long wavelengths is the size of the universe, while at the other end of the spectrum, it is very short and at times devastating.

Just as clean water can become polluted when it travels through a contaminated environment, although it can be initially contaminated at its origin, the same is true when electricity is generated.

Electricity becomes increasingly polluted when it comes into contact with assorted types of electronic equipment. Regular, or "clean," electricity enters buildings at a frequency of 50/60 Hz, although it might already be contaminated; power becomes "dirty" or more polluted when it develops scattered higher-frequency signals as a result of contact with equipment such as computers, plasma televisions, and some appliances. Nonionizing radiation generated by dirty power may radiate to contaminate the adjacent environment and is alleged to be potentially harmful.[12]

Healthy Lighting

According to Rea, M.S. and M.G Figueiro,[13] because the earth rotates on its axis, there is a regular and predictable 24-hour pattern of daylight and darkness over most of its surface. Terrestrial species have adapted to this daily pattern by evolving biological rhythms that repeat at approximately 24-hour intervals.[14] These rhythms are called circadian rhythms, from the Latin circa (about) and dies (day), and reflect a very tight coupling between the natural environmental light-dark pattern and the cyclicity of the endogenous master clock, the suprachiasmatic nuclei (SCN), a small portion of the hypothalamus of the brain. The master clock governs a wide range of biological cycles, from cell division, to hormone production, to behavior (e.g., sleep-wake) that, when synchronized with the natural light/dark cycle, enables the organism to entrain these cycles to its particular photic niche (diurnal or nocturnal) and its location on earth. Although the period of the solar cycle is almost exactly 24 hours, the period of the internal, genetically self-governing master clock is slightly longer than 24 hours in most diurnal species and slightly shorter in most nocturnal species.

A wide variety of anatomical and physiological, as well as behavioral (night-active vs day-active), characteristics are, not coincidentally, associated with the intrinsic period of the master clock,[15] including the retinal mechanisms that support both the visual and circadian systems of nocturnal and diurnal species. For example, the circadian phototransduction mechanisms for nocturnal species are nearly 3000 times more sensitive to light than those of diurnal species.[16] Thus, very low levels of light can entrain the circadian system of a mouse, whereas much higher levels are needed for humans.[17] In this way, humans can avoid false positive light input to the circadian system during the night (e.g., firelight) and mice can avoid false-negative light input at early dawn and late twilight. Furthermore, since the retinal photopigments underlying circadian phototransduction differ between species, light as it affects the circadian system of a nocturnal animal, like a mouse, is quite different than light as it affects the circadian system of humans.[17,18] Mice, for example, have photopigment maximally sensitive to ultraviolet radiation, prevalent in the sky during dawn and dusk, whereas humans are insensitive to ultraviolet radiation. Within species, too, the neural machinery of the retina needed to process light for the visual system and the circadian system are quite distinct, even though the two systems share at least some of the same retinal photopigments.[17,18] In humans, for example, the circadian system receives input from the short-wavelength (S) cone and, consequently, is maximally sensitive to radiation close to this photoreceptor's peak spectral sensitivity, which is about 450 nm. Although the human visual system receives input from all three cone types, the visual channel used to read, for example, receives minimal input from the S cone and is maximally sensitive to middle-wavelength (555 nm) light, representing the combined input of the long- (L) and medium-wavelength (M) cones. Since phototransduction mechanisms differ both among and within species, the very definition of light as a photic stimulus also differs both among and within species.

According to Rea, M.S. and M.G Figueiro,[13] the goal is to elucidate more clearly the differences between light as a stimulus for the human visual system and for the human circadian system. *Circadian light cannot be defined in terms of current photometric quantities* (lux, lumens, luminance) because the phototransduction mechanisms associated with vision, which underlie conventional photometry, are different from those associated with the circadian system. A second goal is to provide a deeper appreciation for the significance of the biological consequences of disrupting circadian rhythms as they might be affected by the electric and natural light sources that illuminate daytime and nighttime built environments. The need for this understanding is becoming increasingly clear as more clinical data demonstrate that disruption of a regular, 24-hour pattern of light and dark leads to wide variety of health- and performance-related problems in humans. Often, the chemically sensitive and chronic degenerative disease patient has disruption in the light-dark cycle. This disturbs their well-being, often giving not only brain dysfunction but also exacerbating weakness and fatigue, as well as increasing chemical sensitivity. To begin this discussion, it is important to first have a brief overview of the phototransduction mechanisms to and the output rhythms from the circadian system.

Input Pathways to the Circadian System

According to Rea, M.S. and M.G Figueiro,[13] photoreceptors in the mammalian retina convert light into neural signals sent via the retinohypothalamic tract (RHT) to the master clock in the SCN. The well-known visual photoreceptors, rods (operating at low, scotopic light levels) and cones (operating at high, photopic light levels) and a recently discovered intrinsically photosensitive retinal ganglion cell (ipRGC)[19] all participate in the conversion of retinal light exposures into neural signals for the SCN, a phenomenon called circadian phototransduction.[20]

The ipRGC neurons directly convert light into neural signals in mammals and, as shown experimentally, can still provide light/dark entraining information to the SCN in the absence of functional rods and cones.[21] Normally, ipRGCs in humans receive input from the cones through a spectrally opponent mechanism that underlies human color vision.[17,22,23] The threshold for participation in circadian phototransduction by this spectrally opponent mechanism appears to be controlled by a neural mechanism in the retina similar to the one that determines whether rods or cones will provide scotopic or photopic input to the visual system.

There are two spectrally opponent cone mechanisms underlying trichromatic human color vision: a blue-vs-yellow channel and a red-vs-green channel. Signals from the three cone photoreceptor types are combined by postreceptor retinal neurons to code color vision. Within the retina, these signals combine in an antagonistic way to form the red-vs-green (L – M) and blue-vs-yellow [S – (L + M)] spectrally opponent channels. These spectrally opponent channels combine with the achromatic channel (primarily L and M) to produce our perceptions of brightness. Because of the spectrally opponent input, the apparent brightness of lights of different spectral power distributions exhibits what is called *subadditivity*, whereby under certain conditions, adding more light can actually reduce the brightness response of the visual system. Subadditivity by the visual system is demonstrated by comparing the apparent brightness of polychromatic lights to the apparent brightness of each narrow-band spectral component. Under some conditions, the combination of two narrow-band spectra together will appear less bright than either of the two narrow-band spectra alone.[23]

Subadditivity is also exhibited by the human circadian system because it receives input from at least one of these spectrally opponent color channels. The blue-vs-yellow channel seems to provide input to the human ipRGC, giving the human circadian system a peak spectral sensitivity at about 450 nm and a subadditive response to white light.[24]

Although much more needs to be discovered about the neural mechanisms underlying circadian phototransduction, input from the blue-vs-yellow channel appears to be controlled by a neural mechanism in the retina similar to the one that transitions the visual system from the scotopic (rod) to the photopic (cone) operating ranges. A model of circadian phototransduction has been developed and used to predict the effectiveness of both narrow-band and polychromatic lights for stimulating the human circadian system, and independent data were subsequently gathered that are consistent with predictions from the model.[23,24]

Some Output Rhythms Affected by the Circadian System

According to Rea, M.S. and M.G Figueiro,[13] *output rhythms are behavioral and physiological rhythms regulated by the SCN, such as sleep, alertness, core body temperature, locomotor activity, feeding and drinking, and hormone production.* A regular 24-hour pattern of light and dark will entrain and orchestrate these rhythms under normal conditions.

Sleep/Wake Cycle

The sleep/wake cycle is one of the most obvious circadian rhythms and defines a species' photic niche; diurnal species are awake during the day and asleep at night, whereas nocturnal species are asleep during the day and awake at night. Most species, including humans, are under the influence of two systems: the *sleep drive* (homeostatic) and the *alerting force* (circadian).[25] The sleep drive and alerting force are distinct and independent from each other, although they normally work in concert to ensure that individuals fall asleep at night and are awake during the day. The *sleep drive* is normally low when people get up in the morning and increases steadily throughout the waking day.

The *alerting force* is regulated by the SCN and thus follows a strong circadian rhythm, reaching a peak during the early evening and a trough during the second half of the night.

Under normal conditions, the close orchestration of the timing of the sleep drive and alerting force determines when individuals fall asleep and how well they sleep at night. *Because the sleep/ wake cycle, and more general behavior, is under the control of two distinct mechanisms, is it not an ideal marker for circadian system response.*

People can be awake and active when their alerting force is low and asleep when it is high. Other markers of circadian timing, like core body temperature and melatonin secretion, are often preferred to behavioral measures, like the sleep/wake cycle, because they are less susceptible to what is termed "masking," whereby extraneous factors compromise valid inferences about circadian timing. Unfortunately, there is a subset of chemically sensitive patients who have their sleep-wake cycle distorted, and some can sleep only 2 hours per night, which markedly exacerbates their chemical and/ or electrical sensitivity, preventing them from getting well.

Core Body Temperature

Core body temperature is governed by the hypothalamus and varies with a circadian pattern. Temperatures are high during the day, reach a peak in the early evening, and then decline to a nadir about 2 hours before one naturally wakes up. *Core body temperature is typically in synchrony with the alerting force from the SCN.* Therefore, core body temperature is often used as a primary measure of circadian timing. Although core body temperature is less susceptible to masking, it is not impervious to other extraneous variables such as digestion and exercise. Core body temperature readings are also difficult to obtain because they are typically acquired through rectal probes. Controversy also exists as to whether core body temperatures gathered by this technique are really indicative of the intercranial temperatures that are most closely regulated by the hypothalamus. The chemically and electrically sensitive are frequently cold sensitive. The temperature runs from 96.5°F to 98°F. Some of this may be due to vascular spasm, but some may be due to the temperature stat in the brain.

Hormone Production

Although there is strong evidence that the master clock influences the circadian rhythms of a wide variety of hormones, most measures of circadian timing are based upon concentration measurements of the hormone melatonin synthesized by the pineal gland. The pineal gland is located near the center of the brain, is about the size of a pea, and has a shape resembling that of a pine cone, thus its name. It is believed that the primary function of the pineal gland is to convey light/dark information to the body via nighttime secretions of the hormone melatonin.[26] Melatonin is easily absorbed into the bloodstream, which makes it an ideal chemical messenger of time-of-day information to the entire body. Melatonin also participates in the transmission of information concerning day length or photoperiod for the organization of seasonal responses in some species (e.g., breeding). The chemically sensitive patient often has mild to profound changes in the different seasons. Often they are better in the summer when there is more light and worse in the winter when there is less light.

Melatonin is only produced at night *and* under conditions of darkness.[25] Under normal conditions, changes in melatonin concentrations are approximately inversely related to changes in core body temperature in humans; peak levels typically occur at night slightly before core body temperature troughs. Melatonin concentrations, as well as those of other hormones such as cortisol, are measured in a wide variety of body fluids; blood plasma and saliva are the two most practical sources for melatonin concentration measurements. Unlike other hormones, however, melatonin appears to be almost exclusively regulated by the SCN. Since the response of the SCN is strongly affected by retinal light exposures, the (masking) effect of light on melatonin synthesis at night has been used as the primary measure of the circadian system's response to light. It is now well established that modulation of light level and light spectrum will induce dose-response, graded changes in melatonin concentrations at night. For practical and theoretical reasons, then, most of the discussion in the literature about the impact of light on the circadian system is based upon suppression of melatonin production by the pineal gland at night.

Light as a Physical Quantity

According to Rea, M.S. and M.G Figueiro,[13] perhaps surprisingly, light is narrowly and formally defined in terms of human vision, and its definition is based upon psychophysical experiments investigating the spectral sensitivity of the eye that were conducted nearly a century ago.[27,28] Strictly speaking, then, light has no meaning for plants and other animals. Nevertheless, light is commonly used to describe any and all optical radiation between approximately 380 and 730 nm, even though, for example, some plants and animals have sensitivity to optical radiation outside this region. People often erroneously use adjectives to expand this spectral range with such terms as "ultraviolet light" or "infrared light," both of which are oxymorons. *Since neither ultraviolet nor infrared radiation can be seen by humans, optical radiation in these bands cannot be light, by definition.*

Even for humans, the formal definition of light is quite limiting. The human visual system is very complex and will exhibit a multitude of spectral sensitivity functions depending upon a range of spectral-spatial-temporal factors associated with that optical radiation. For example, the apparent brightness of a small blue disc of light seen by the fovea will appear dimmer than that same disc seen in the periphery.[29] Over the last century, a large number of human visual spectral sensitivity functions have been identified. To at least in part address this complexity, a handful of spectral sensitivity functions have been formally accepted as luminous efficiency functions for optical radiation and, perhaps surprisingly, as alternative definitions for light.[27,30,31] The currently accepted luminous efficiency functions that underlie alternative, formal definitions of light, which can lead, and have led, to significant confusion among nonspecialists. The modeled spectral sensitivity of the human circadian system to polychromatic white light.

According to Rea, M.S. and M.G Figueiro,[13] V(λ) is the oldest of these luminous efficiency functions and underlies the definition of light used in commerce and by all government regulatory agencies. Thus, when a lamp manufacturer publishes the number of *lumens* generated by a source, V(λ) is used to spectrally weight its optical radiation. When government agencies regulate the *luminous efficacy* of a light source, the *illuminance* needed for an application (e.g., a roadway intersection), or the *luminous intensity* of a signal light, V(λ) is used to spectrally weight the optical radiation from practical light sources in all of these regulations. Significantly, then, when one purchases a light source or is legally constrained in the application of that source, V(λ) is embedded in every lighting specification.

According to Rea, M.S. and M.G Figueiro,[13] because the formal definition of light does not represent the spectral sensitivities of plants, animals, and, indeed, the complexity of the human visual system, V(λ) is an incomplete and potentially misleading characterization of optical radiation from sources used to grow plants and animals and even to provide visual sensations to humans. To add further weight to this problem, and as already discussed, light is necessary to control the daily cycles of all human biological functions.[22] *Circadian rhythms are governed by retinal phototransduction mechanisms quite distinct from those underlying V(λ).* Because science and medicine have increasingly shown the importance of circadian rhythms for health and well-being in humans, it is likewise increasingly important for manufacturers and regulatory bodies to expand their definitions of light to include spectrally weighted optical radiation that entrains the human circadian system.

In order to fully understand the importance that light has on the biological system, it is first necessary to consider optical radiation in terms of its fundamental physical properties. For the purposes of this chapter and for practical convenience, light can be decomposed into five basic characteristics:[32,33] quantity, spectrum, spatial distribution. timing, and duration.

Light Characteristics as They Affect the Human Visual and Circadian Systems

As previously noted, the natural 24-hour light/dark cycle is the master clock's main synchronizer for its host's photic niche and location on earth. To more precisely understand how this synchronization occurs, it is necessary to understand the physical characteristics of light incident on the retina and, in these terms, how light is processed by the host's retina. Because light is formally defined in terms of the human visual system,[30] and because the human species is of primary interest for lighting technology development, it is useful to contrast how each characteristic of light is processed by the human visual system and the human circadian system.

Quantity

The human visual system can operate over the entire very large range of levels of light available on the earth's surface, from starlight (0.00001 lx) to bright sunshine on a snow-covered mountain (100,000 lx), about 10–12 orders of magnitude. Typical levels provided by electric light sources, both indoors and out, are in the middle of this range, from about 0.1 lx to 1000 lx. Although the visual system operates better at higher levels, it does so with a diminishing return. Several studies (e.g., [34]) have shown, for example, that visual performance, the speed and accuracy of processing visual information, is near its maximum at typical office illuminance levels (500 lx on the task surface or approximately 100 lx at the cornea) using commercially available "white" light sources. In contrast, these same levels and sources provide light only near the threshold of circadian system activation which is approximately 30 lx at the cornea with natural pupils ($\approx$4 mm diameter) from a white light source.

Spectrum

According to Rea, M.S. and M.G Figueiro,[13] visual performance is most efficient for light in the middle of the visible spectrum, with a spectral sensitivity closely matching the photopic luminous efficiency function, peaking at 555 nm. The human circadian system, on the other hand, is most sensitive to much shorter wavelengths, near about 450 nm[24] and, as previously discussed, exhibits subadditivity. The circadian system spectral sensitivity has a negative lobe from about 550 to 700 nm.[17] When spectral energy in this long wavelength range is combined with spectral energy at short wavelengths, the net impact of the short-wavelength energy on the circadian system response is reduced. In general, then, different practical light sources with the same efficiency for visual performance can vary quite considerably in terms of their efficiency for stimulating the circadian system, and vice versa.

As previously described, a general model of circadian light from any spectral power distribution and irradiance has been published,[22,24] and it is possible to compare practical light sources in terms of the efficacy for stimulating the circadian system. Table 1.3 shows the relative amount of corneal (photopic) illuminance needed to produce a criterion response by the human circadian system (50%

TABLE 1.3   Illuminance and Relative Electrical Power to Achieve 50% Melatonin Suppression

Light Source	Illuminance (lux)	Relative Electric Power
2700 K compact fluorescent (Greenlite15WELS-M)	722	18
2856 K incandescent A lamp	511	75
3350 K linear fluorescent (GE F32T8 SP35)	501	10
4100 K linear fluorescent (GE F32T8 SP41)	708	14
5200 K LED phosphor white (Luxeon Star)	515	15
6220 K linear fluorescent (Philips Colortone 75)	349	12
8000 K Lumilux Skywhite fluorescent (OSI)	266	6
Blue LED (Luxeon Rebel, λmax = 470 nm)	30	1
Daylight (CIE D65)	270	NA

Note: Several practical light sources with the photopic illuminance (lm/m^2) at the cornea required to achieve 50% melatonin suppression after 1 hour exposure. The relative electric power (W) need to achieve this same criterion response, based upon the current technologies, is also listed. Table 1.3 assumes 50% melatonin suppression for 1 hour exposure with a fixed pupil diameter of 2.3 mm.

melatonin suppression for 1 hour exposure with a pupil diameter of 2.3 mm). In general, the more the source is dominated by short wavelengths, the lower the illuminance required to reach a criterion response; compare, for example, two "white" fluorescent lamps of different correlated color temperatures (CCTs), the 8000-K lamp and the 2700-K compact fluorescent lamp. For these two examples, it takes nearly three times the illuminance and electric power to produce the same circadian response. Not unexpectedly, then, the blue LED is the most efficacious light source for stimulating the human circadian system. One peculiarity should be noted in the table. There is an intransitivity between CCT and circadian effectiveness, as shown by the higher required illuminance for the 4100 K lamp compared to the 3350 K lamp. This intransitivity is associated with the subadditive response of the circadian system according to the model by Rea, M.S. et al.[24]

The relative effectiveness of different light sources for stimulating the circadian system plotted as a function of photopic illuminance at the cornea. Also shown is a relative visual performance (RVP) response function.[21] Since the spectral sensitivity of the visual system as measured by RVP is well described by the photopic luminous efficiency function, the single RVP function characterizes visual performance for "white" light sources of different spectral irradiance functions, like those in Table 1.3.

Spatial Distribution

According to Rea, M.S. and M.G Figueiro,[13] the effectiveness of light for the human circadian system cannot be determined from the spectral irradiance distribution alone. In contrast to the fundamental significance of optical refraction for image formation by the cornea and crystalline lens for the visual system, the circadian system appears to care little about the distribution of light across the retina. Although there is some evidence that the spatial sensitivity of the circadian system is not isotropic across the entire retina,[35] it would seem that, as a first approximation, the circadian system responds simply to the total amount of circadian light incident on the retina. Light incident on the cornea will be, depending upon angle, differentially effective for entering the pupil and thereby stimulating the retinal mechanisms necessary for stimulating the circadian system. Using a ray-tracing technique, Van Derlofske et al.,[36] developed a spatial model of retinal irradiance that depends upon the angle of light incident on the cornea as occluded by the brow, cheek, and nose.[36] Naturally, these complications vary with each person, but because they are limited to large entrance angles, the spatial sensitivity of the retina appears to be closely approximated by a *three-dimensional cosine distribution*. According to this simpler model, the effectiveness of light incident on the cornea at increasingly eccentric angles falls off as the cosine from the optical axis of the pupil.

Although this simple model is a useful first approximation of the spatial sensitivity of the circadian system, it will be increasingly important for the development of practical applications of circadian light that the distribution of neural phototransduction mechanisms underlying the circadian system in the retina be better understood. For example, if the circadian system is more sensitive to light incident on the lower retina than the upper retina, as one study suggests,[37] then lighting systems will need to be designed to deliver light to the cornea from above the normal viewing angle (e.g., an LED source mounted just above the computer screen).

Duration

According to Rea, M.S. and M.G Figueiro,[13] the *amount of time an individual is exposed to a light source must also be specified in order to predict the effect of light on the circadian system.* Because the visual system is a remote sensing system and needs to respond to threats and opportunities very quickly, it will respond within a few hundred milliseconds to a briefly flashed or moving target. In contrast, the circadian system appears to be quite slow to respond, if at all, to short-duration flashes (like lightning flashes) and, apparently, does not respond at all to moving light stimuli because of its coarse spatial sensitivity. Although little is known about the circadian system phase-shifting response to flashes of variable durations, the suppression of melatonin content in the bloodstream begins to respond after approximately 2 to 10 minutes of sufficient light exposure. Based on studies conducted by McIntyre and colleagues,[38] the amount of time required to achieve a criterion response by the circadian system will depend upon the duration of the light exposure. For example, 50% melatonin suppression by a "white" light (5500 K) at night took about 28 minutes following exposure to 3000 lx at the eye and about 33 minutes following exposure to 1000 lx at the eye, but 50% suppression could not be achieved following a 100 lx exposure level for any duration. The duration of light exposure needed to measure 50% nocturnal melatonin suppression by humans.[38]

From a practical perspective, provision must be made for prolonged exposures to light (and dark) for the circadian system to synchronize its operation to a 24-hour day.

Timing

Although the mechanisms are still unclear, the minimum amount of light needed for detection by the visual system changes only slightly over the 24-hour day.[39] Despite these slight changes, the visual system is essentially "ready to go" at any time of the day or night. In stark contrast, the response of the circadian system will change in the magnitude and direction of response depending on the timing of light exposure. Light can have a large or a small effect on the master clock and can either phase advance or phase delay the clock depending upon *when* during the circadian cycle the retina was exposed to light.[40,41] Light-induced phase advances reset the master clock to an earlier time, whereas delays will reset the clock to a later time. For example, if an otherwise synchronized person is exposed to light late at night, the master clock will tell the person to go to bed earlier and wake up earlier the following day. However, if that same person is exposed to light in the early evening, the clock will send information to go to bed later and wake up later the following day.

Because the typical period of the human biological clock is a bit longer than 24 hours, light in the morning upon awakening will advance the clock's timing every morning in order to keep biological functions synchronized with the solar day. From a practical perspective, then, *sufficient morning light (quantity, spectrum, spatial distribution, and duration) helps humans synchronize their lives to their diurnal schedules.* The same evening light exposure will, however, delay bed and rise times, potentially limiting sleep and compromising well-being. Understanding the timing of circadian light exposure is perhaps the most important characteristic to consider when developing new lighting technologies and applications.

History of Light Exposure on Circadian Systems

Finally, a word should be said about light history. Light exposures on preceding days and weeks influence the circadian system's current sensitivity to light. Rea, M.S. et al.'s study[24] demonstrated that, depending on the amount of light to which a laboratory animal was exposed in one part of the 24-hour day, dim light given in the other part of the day was interpreted by the circadian system as either light or dark.[42] If the animal was alternately exposed to bright and dim light for 12 hours each, the dim light was interpreted as the dark half of the light-dark cycle. If, however, the same animals were alternately exposed to the dim light and darkness for 12 hours, the same dim light was interpreted as the light half of the light-dark cycle. In humans, the higher the exposure to light during the day, the lower the sensitivity of the circadian system to electric lights at night (e.g., Hebert et al.[43]). These findings in both animals and humans demonstrate that the circadian system responds to *changes* in the 24-hour light-dark cycle with less regard for the absolute levels.

From a practical perspective, then, knowing an individual's history of light exposure is important for predicting the impact of light on, say, a farmer versus a computer programmer. Different lighting schemes may therefore have to be developed for the workplace and the home depending upon the individual's overall lifestyle.

Consequences and Implications

The human endogenous master clock has evolved to recognize a regularly occurring, 24-hour pattern of light and dark. There are, of course, genetic differences among individuals, so the coupling between the highly predictable 24-hour solar day and one's own internal clock will differ to some degree. For example, individuals can be characterized by what is termed their chronotype;[44] some people are *larks*, rising early in the day and going to bed early at night, while others are *owls*, going to bed late and, if they can, rising late in the morning. Interestingly, as individuals reach puberty, the propensity to become more owl-like increases until, when one reaches young adulthood, he or she returns to a more "normal" pattern of sleep-wake.[45] Toward the end of an individual's life, apoptosis begins to become problematic for many physiological systems, including circadian regulation. One begins to lose retinal neurons, compromising not only the ability to see, but also the ability to register the robust daily pattern of light-dark that is needed to maintain entrainment. Individuals also lose neurons in the master clock itself, particularly in persons with Alzheimer's disease (AD),[46,47] further compromising our ability to synchronize both physiology and behavior to the 24-hour day.

Humans are a diurnal species, typically awake during the day and asleep at night. Thus, most vocations are performed during daylight hours, but nearly 15% of the US population is considered shift workers, whereby they are active during the night and asleep during the day.[48] These people are rarely, however, true night-shift workers. Rather, most shift workers live a compromise between participation in a day-active society (friends, families, bank hours) and night-active employment. These people rarely have a regular, 24-hour light dark cycle. Epidemiological evidence indicates rather strongly that shift workers are more susceptible to obesity,[49] cardiovascular disease,[50] and cancers.[51]

As will be discussed, seniors are the population most likely to be positively affected by a prescribed 24-hour light/dark pattern. It is much more difficult to prescribe a light/dark pattern that will significantly improve sleep efficiency and alertness in rotating-shift workers. Little is known about the endogenous rhythms of rotating-shift workers, but, more importantly, by definition for rotating shifts, they are going to experience aperiodic light/dark exposures. Consequently, rotating-shift workers are almost certainly not entrained to their work schedules and, as a result, experience symptoms similar to those experiencing "jet-lag" following transoceanic flights. Although high levels of light exposure during the night shift can increase acute alertness and reduce feelings of sleepiness,[51,52] disruption of a regular 24-hour pattern of light and dark compromises entrainment. Moreover, acute suppression of nocturnal melatonin by high light levels at night has been linked to increased risk of cancer.[53] Also, melatonin is an antioxidant that is eight times more potent than any other antioxidant in the body; therefore, it is important to preserve its function and circadian rhythm. People with chemical sensitivity often are sensitive to melatonin. They have melatonin neutralized by the intradermal neutralizing technique. However, this procedure often doesn't work and then melatonin can't be used to induce sleep. Therefore, it may simply be impossible to develop a prescribed lighting scheme for buildings where people work rotating shifts. It may nevertheless be possible to help minimize the problems associated with rotating-shift workers (poor sleep, poor performance, gastrointestinal disorders, obesity, and breast cancer) by continuously controlling light/dark exposures on an individual basis. The science necessary for developing this interventional approach is just beginning, however, so no practical means of continuously controlling personal light/dark exposures is currently available for this purpose.

The combinations of genetic and environmental factors influencing circadian entrainment are infinite, so it is quite difficult to ascertain risk factors for a given individual and, moreover, to prescribe a "recipe" for minimal risk of disease on a personal level. It is possible, however, to identify populations who might be at risk and those populations where interventions would most likely be useful. The less that is known about the endogenous factors associated with circadian rhythms in a population or the less that is known about their exposures to light and to dark, the less likely it will be that successful corrective interventions will be implemented.

Total and Cause-Specific Mortality of U.S. Nurses Working Rotating Night Shifts

Shift work including a nighttime rotation is common, with up to 20% of the Western workforce encountering alternative work schedules some time during their career.[54] Observational studies[55-71] consistently suggest positive associations of night-shift work and cardiovascular disease (CVD)[55] and cancer[56-71] risk. Thus, the WHO[72] in 2007 classified night work as a probable carcinogen. In addition to the link with cancer and CVD risk, other health outcomes, including diabetes, hypertension, chronic fatigue, various sleep problems,[73-76] and higher body weight,[75] are associated with shift work.[77-79] However, evidence is more limited and inconsistent[80] on associations between night-shift work and all-cause and cause-specific mortality in observational studies.[71,81-85]

The circadian system and its prime marker, melatonin, are considered to have antitumor effects through multiple pathways, including antioxidant activity, anti-inflammatory effects, and immune enhancement.[86-89] They also exhibit beneficial actions on cardiovascular health by enhancing endothelial function, maintaining metabolic homeostasis,[90] and reducing inflammation.[91,92] Direct nocturnal light exposure suppresses melatonin production[93] and resets the timing of the circadian clock.[86,94,95] In addition, sleep disruption may also accentuate the negative effects of night work on health.[95,96] Taken together, substantial biological evidence supports the role of night-shift work in the development of poor health conditions, including cancer, CVD, and ultimately, mortality.

The current analysis is based on 22 years of follow-up of 74,862 women participating in the Nurses' Health Study (NHS). This large U.S. cohort of nurses provides a unique opportunity to study associations between duration of rotating night-shift work, all-cause, and cause-specific mortality. The baseline age range (30–55 years), high death rate, and nearly complete follow-up allows for a powerful analysis of night-shift work and mortality, with repeat measures available on a number of important confounding factors. The 1988 assessment of lifetime night work likely captures most of the night-shift work history of women retained in this cohort, as, based on data from the NHS2 cohort (a similar cohort, which comprised 116,434 younger women, all registered U.S. nurses),[97] <5% of nurses continue to work night shifts in middle age and <2% commence working night shifts at this age.

Results

Over 22 years and 1.5 million person-years of follow-up, 14,181 deaths were documented, of which 3062 were attributed to CVD and 5413 to cancer. Women who had never worked rotating night shifts accounted for 41% of the person-years of follow up, and those who worked 1–5, 6–4, and ≥15 years on shifts accounted for 41%, 11%, and 7% of person-years, respectively. Table 1.4 shows that women with longer durations of rotating night-shift work tended to be older (mean age 66.3 years for ≥15 years of shift work vs 63.6–64.6 years for others), had a higher BMI, and were more physically active after standardizing for age. They were also more likely to be current smokers, less likely to use postmenopausal hormones or multivitamins, and, of those married, the husbands tended to be less educated. There were no appreciable differences in dietary factors across durations of shift work; however, women who worked night shifts for longer durations tended to drink less alcohol and ate less daily cereal fiber compared to women without night-shift work. They also had gained more weight since

TABLE 1.4 Age-Standardized Characteristics of the NHS Population throughout the Follow-Up Period 1992–2008

Characteristic	Study Population
No. of women	115,745
Age (years) (mean ± SD)	65.9 ± 8.3
White race/ethnicity (%)	96.8
BMI (kg/m^2) (mean ± SD)	25.6 ± 7.4
Smoking status (%)	
Never	43.8
Past	43.4
Current	12.6
Missing	0.2
Pack-years of smoking[a] (mean ± SD)	24.4 ± 21.5
High-risk diet score (mean ± SD)	164.3 ± 100.4
Total caloric intake (kcal/day) (mean ± SD)	1515.1 ± 724.7
Physical activity (%)	
<3 MET hr/week	20.6
3 to <9 MET hr/week	20.7
9 to <18 MET hr/week	18.5
18 to <27 MET hr/week	11.4
>27 MET hr/week	19.3
Missing	9.4
Husband's education attainment (%)	
Less than high school	4.0
High school	25.8
Greater than high school	35.2
Missing	35.0
Parity (%)	
Nulliparous	5.8
1	7.3
2–3	55.0
≥4	29.9
Missing	2.1
Menopausal status (%)	
Premenopausal	5.1
Postmenopausal and no HRT use	22.2
Postmenopausal and past HRT use	28.8

(Continued)

TABLE 1.4 (Continued)   Age-Standardized Characteristics of the NHS
Population throughout the Follow-Up Period 1992–2008

Characteristic	Study Population
Postmenopausal and estrogen replacement therapy use	14.0
Postmenopausal and estrogen-progesterone replacement therapy use	9.5
Missing	20.4
Comorbidities	
Hypertension (%)	46.7
Coronary heart disease (%)	3.0
Rheumatologic disease (%)	3.8
Nonaspirin nonsteroidal anti-inflammatory drug use (%)	22.1
Warfarin use (%)	2.2
Multivitamin supplement use (%)	45.2
Median family income[b] (US$1000s) (mean ± SD)	63.7 ± 24.7
Median house value[b] (US$10,000s) (mean ± SD)	17.1 ± 12.7
Predicted $PM_{2.5}$ ($\mu g/m^3$) (mean ± SD)[c]	12.6 ± 3.0
Interquartile range	4.1
Predicted PM_{10} ($\mu g/m^3$) (mean ± SD[c]	20.8 ± 5.8
Interquartile range	6.4
Predicted $PM_{2.5-10}$ ($\mu g/m^3$) (mean ± SD)[c]	8.2 ± 4.2
Interquartile range	4.6
Distance to road [m (%)]	
0–49	15.1
50–199	27.6
200–499	32.1
≥500	25.1

Source: U.S. Census Bureau. 2000. *Environmental Health Perspectives* 123(12): 1627.[141]

Abbreviations: HRT, hormone replacement therapy; MET, metabolic equivalent.

[a] Among smokers.

[b] Estimated for census tract of residence using data from the U.S. Census Bureau (2000).

[c] For 12 months prior.

age 18 years, and were more likely to have a history of diabetes, hypertension, and hypercholesterolemia.

Age- and multivariable-adjusted HRs for the associations between night-shift work and all-cause and cause-specific mortality are presented in Table 1.5. Women working rotating night shifts for >5 years had modest increases in both all-cause and CVD-related mortality compared to women who never worked rotating night shifts. For all-cause mortality, the multivariable-adjusted hazard ratios (HRs) for women with 1–5, 6–14, or ≥15 years of rotating night-shift work were, respectively, 1.01 (95% CI = 0.97, 1.05), 1.11 (95% CI = 1.06, 1.17), and 1.11 (95% CI = 1.05, 1.18; $_{Ptrend}$ < 0.001). For CVD mortality, the multivariable adjusted HRs were, respectively, 1.02 (95% CI = 0.94, 1.11), 1.19 (95% CI = 1.07, 1.33), and 1.23 (95% CI = 1.09, 1.38; $_{Ptrend}$ < 0.001). Women working night shifts for 6–14 or ≥15 years also had a higher risk of all-cancer mortality.

Compared to women who never worked rotating night shifts, the age-adjusted HRs were, respectively, 1.10 (95% CI = 1.01, 1.20) and 1.20 (95% CI = 1.09, 1.32), but these associations were attenuated after multivariable adjustment, with HRs of 1.04 (95% CI = 0.95, 1.13) and 1.08 (95% CI = 0.98, 1.19), respectively.

Results for specific cancer sites with ≥200 deaths are offered in Table 1.6. For lung cancer mortality, the age-adjusted HR for ≥15 years of rotating night-shift work was 1.44 (95% CI = 1.20, 1.73) and 1.25 after multivariate adjustment (95% CI = 1.04, 1.51), with smoking as the driving confounder. For colorectal cancer, the age-adjusted HR after ≥15 years of rotating night-shift work was 1.42 (95% CI = 1.04, 1.94), but was attenuated to 1.33 (95% CI = 0.97, 1.83) after multivariate adjustment. Adjusting for current disease status (T2D, hypertension, and hypercholesterolemia) left results largely unchanged (data not shown). Although no increase in breast cancer mortality was observed among women with ≥15 years of rotating night-shift work (multivariable HR = 0.99, 95% CI = 0.74, 1.33), among women with ≥30 years of rotating night-shift work, the multivariable HR for breast cancer mortality was 1.47 (95% CI = 0.94, 2.32). There was a nonsignificant increase in mortality from kidney cancer and myeloma among women with 6–14 and ≥15 years of night work: the multivariable HRs were 1.72 (95% CI = 1.03, 2.86) and 1.39 (95% CI = 0.75, 2.57) for

TABLE 1.5 Association between Night-Shift Work and All-Cause, CVD, and Cancer Mortality in NHS, 1988–2010 (n = 74,862)

Mortality	Never	Night-Shift Work Duration			p for Trend
		1–5 Years	6–14 Years	≥15 Years	
All causes					
No. of deaths	5,417	5,424	1,910	1,430	
Age-adjusted HR (95% CI)	1.00 (ref)	0.97 (0.94, 1.01)	1.19 (1.13, 1.25)	1.24 (1.17, 1.32)	**<0.001**
Multivariable HR (95% CI)[a]	1.00 (ref)	1.01 (0.97, 1.05)	1.11 (1.060, 1.17)	1.11 (1.05, 1.18)	**<0.001**
All cardiovascular disease					
No. of deaths	1,128	1,128	442	364	
Age-adjusted HR (95% CI)	1.00 (ref)	0.97 (0.90, 1.06)	1.30 (1.16, 1.45)	1.45 (1.29, 1.63)	**<0.001**
Multivariable HR (95% CI)[a]	1.00 (ref)	1.02 (0.94, 1.11)	1.19 (1.07, 1.33)	1.23 (1.09, 1.38)	**<0.001**
All cancer					
No. of deaths	2,087	2,148	672	506	
Age-adjusted HR (95% CI)	1.00 (ref)	1.00 (0.94, 1.06)	1.10 (1.01, 1.20)	1.20 (1.09, 1.32)	**<0.001**
Multivariable HR (95% CI)[a]	1.00 (ref)	1.03 (0.97, 1.09)	1.04 (0.95, 113)	1.08 (0.98, 1.19)	0.11

Note: Boldface indicates statistical significance ($p \leq 0.05$).

[a] Multivariable model adjusted for age (continuous), alcohol consumption (none, 0.1–4.9, 5.0–14.9, ≥15.0 g/day), physical exercise (metabolic equivalent values; quintiles), multivitamin use (yes, no), menopausal status (premenopausal, postmenopausal) and post-menopausal hormone use (never, past, and current), physical exam in the past 2 years (no, yes for symptoms, and yes for screenings), healthy eating score (quintiles), smoking status (never, former, and current), pack-years (<10, 10–19, 20–39, ≥40 for former smokers; <25, 25–44, 45–64, ≥65 for current smokers), BMI (<21, 21–22.9, 23–24.9, 25–27.4, 27.5–29.9, 30–34.9 ≥35), and husband's education (less than high school, some high school, high school graduate, college, graduate school, missing category).Abbreviations: CVD, cardiovascular diseases; HR, hazard ratio; NHS, Nurses' Health Study.

TABLE 1.6 Association between Night-Shift Work and Cancer-Specific (≥200 Deaths) Mortality in NHS 1988–2010 (n = 74,862)

Mortality	Never	Night-Shift Work Duration			p for Trend
		1–5 Years	6–14 Years	≥15 Years	
Lung cancer					
No. of deaths	501	523	168	150	
Age-adjusted HR (95% CI)	1.00 (ref)	1.02 (0.91, 1.16)	1.14 (0.96, 1.36)	1.44 (1.20, 1.73)	<0.001
Multivariable HR (95% CI)[a]	1.00 (ref)	1.05 (0.92, 1.19)	0.99 (0.83, 1.18)	1.25 (1.04, 1.51)	0.05
Breast cancer					
No. of deaths	269	293	79	55	
Age-adjusted HR (95% CI)	1.00 (ref)	1.05 (0.89, 1.25)	1.04 (0.81, 1.34)	1.08 (0.80, 1.44)	0.64
Multivariable HR (95% CI)[a]	1.00 (ref)	1.07 (0.90, 1.26)	0.99 (0.76, 1.27)	0.99 (0.74, 1.33)	0.83
Ovarian cancer					
No. of deaths	180	168	47	30	
Age-adjusted HR (95% CI)	1.00 (ref)	0.92 (0.74, 1.13)	0.90 (0.65, 1.24)	0.86 (0.58, 1.27)	0.41
Multivariable HR (95% CI)[a]	1.00 (ref)	0.95 (0.77, 1.18)	0.89 (0.64, 1.23)	0.82 (0.55, 1.22)	0.27
Pancreatic cancer					
No. of deaths	149	173	52	33	
Age-adjusted HR (95% CI)	1.00 (ref)	1.12 (0.90, 1.39)	1.20 (0.88, 1.65)	1.10 (0.75, 1.60)	0.47
Multivariable HR (95% CI)[a]	1.00 (ref)	1.12 (0.90, 1.40)	1.14 (0.83, 1.58)	1.03 (0.70, 1.51)	0.77

(Continued)

TABLE 1.6 (Continued)   Association between Night-Shift Work and Cancer-Specific (≥200 Deaths) Mortality in NHS 1988–2010 (n = 74,862)

Mortality	Never	Night-Shift Work Duration			
		1–5 Years	6–14 Years	≥15 Years	p for Trend
Colorectal cancer					
No. of deaths	180	176	56	52	
Age-adjusted HR (95% CI)	1.00 (ref)	0.95 (0.77, 1.17)	1.07 (0.80, 1.45)	1.42 (1.04, 1.94)	**0.02**
Multivariable HR (95% CI)[a]	1.00 (ref)	0.98 (0.79, 1.21)	1.05 (0.77, 1.42)	1.33 (0.97, 1.83)	0.07
Non-Hodgkin's lymphoma					
No. of deaths	103	89	34	26	
Age-adjusted HR (95% CI)	1.00 (ref)	0.85 (0.66, 1.09)	0.94 (0.65, 1.37)	1.11 (0.74, 1.66)	0.57
Multivariable HR (95% CI)[a]	1.00 (ref)	0.876 (0.67, 1.12)	0.96 (0.66, 1.40)	1.06 (0.71, 1.58)	0.77
Other cancer (ICD = 199)					
No. of deaths	146	145	35	32	
Age-adjusted HR (95% CI)	1.00 (ref)	0.95 (0.76, 1.20)	0.80 (0.55, 1.16)	1.05 (0.72, 1.55)	0.87
Multivariable HR (95% CI)[a]	1.00 (ref)	0.98 (0.77, 1.23)	0.75 (0.52, 1.09)	0.94 (0.63, 1.38)	0.39

Note: Boldface indicates statistical significance ($p < 0.05$).

[a] Multivariable model adjusted for age (continuous), alcohol consumption (none, 0.1–4.9, 5.0–14.9, ≥15.0 g/day), physical exercise (metabolic equivalent values; quintiles) multivitamin use (yes, no), menopausal status (premenopausal, postmenopausal) and postmenopausal hormone use (never, past, and current), physical exam in the past 2 years (no, yes for symptoms, and yes for screenings), healthy eating score (quintiles), smoking status (never, former, and current), pack-years (<10, 10–19, 20–39, ≥40 for former smokers, <25, 25–44, 45–64, ≥65 for current smokers), BMI (<21, 21–22.9, 23–24.9, 25–27.4, 27.5–29.9, 30–34.9 ≥35), and husband's education (less than high school, some high school, high school graduate, college, graduate school, missing category). HR, hazard ratio; NHS, Nurses' Health Study.

kidney cancer ($_{Ptrend}$ = 0.048) and 1.56 (95% CI = 0.93, 2.64) and 1.61 (95% CI = 0.90, 2.88) for myeloma ($_{Ptrend}$ = 0.08). There was no significant association between rotating night-shift work and mortality from other cancers.

For mortality from specific types of CVD with >100 deaths (Table 1.7), increased ischaemic heart disease (IHD) mortality was observed among women who worked rotating night shifts for 6–14 and ≥15 years (HR = 1.23, 95% CI = 1.03, 1.47; 1.34, 95% CI = 1.11, 1.61, respectively; $_{Ptrend}$ < 0.001). Additional adjustment for current disease status (T2D, hypertension, and hypercholesterolemia) only minimally affected these associations (Table 1.7). There were no significant associations among women working rotating night shifts for 6–14 and ≥15 years for cerebrovascular disease or any other CVD.

When stratifying by smoking status (Table 1.8), the association of rotating night-shift work appeared to be stronger among current smokers for all-cause mortality (HR ≥ 15 years, 1.21 vs 1.11) but remained statistically significant among newer-smokers ($_{Ptrend}$ = 0.05). Similarly, for CVD and IHD mortality, HRs were comparable among newer-smokers and the entire analytic sample. Associations for overall cancer, lung cancer, and colon cancer mortality appeared to vary by smoking status, but there was no statistically significant interaction ($_{Pinteraction}$ > 0.05). No significant interaction with BMI was observed for any of the endpoints (Table 1.8).

In this large prospective U.S. cohort of nurses, working rotating night shifts for >5 years was associated with a significant increase in all-cause and all-CVD (particularly IHD) mortality. Working ≥15 years of rotating night-shift work was associated with a significant increase in lung cancer mortality. There was also a nonsignificant increase in mortality due to cancer overall, and in several specific cancer sites.

Previous studies of shift work and IHD incidence or mortality varied by design and quality, with inconsistent results.[81,98–100] However, results from two prospective follow-up studies[101,102]

TABLE 1.7 Association between Night-Shift Work and CVD-Specific (>100 Deaths) Mortality in NHS, 1988–2010 (n = 74.862)

			Night-Shift Work Duration		
Mortality	Never	1–5 Years	6–14 Years	≥15 Years	p for Trend
Ischemic heart disease					
No. of deaths	407	386	170	153	
Age-adjusted HR (95% CI)	1.00 (ref)	0.92 (0.80, 1.05)	1.38 (1.15, 1.65)	1.69 (1.40, 2.04)	<0.001
Multivariable HR (95% CI)[a]	1.00 (ref)	0.97 (0.85, 1.12)	1.23 (1.03, 1.47)	1.34 (1.11, 1.61)	<0.001
Multivariable HR (95% CI)[b]	1.00 (ref)	0.96 (0.83, 1.11)	1.17 (0.98, 1.41)	1.23 (1.02, 1.49)	0.006
Cerebrovascular disease					
No. of deaths	319	320	119	90	
Age-adjusted HR (95% CI)	1.00 (ref)	0.97 (0.83, 1.14)	1.24 (1.00, 1.53)	1.23 (0.97, 1.55)	0.02
Multivariable HR (95% CI)[a]	1.00 (ref)	1.02 (0.87, 1.20)	1.20 (0.97, 1.48)	1.12 (0.88, 1.42)	0.17
Multivariable HR (95% CI)[b]	1.00 (ref)	1.01 (0.86, 1.18)	1.16 (0.94, 1.44)	1.06 (0.83, 1.340)	0.36

(Continued)

TABLE 1.7 (Continued)   Association between Night-Shift Work and CVD-Specific (>100 Deaths) Mortality in NHS, 1988–2010 (n = 74.862)

		Night-Shift Work Duration			
Mortality	**Never**	**1–5 Years**	**6–14 Years**	**≥15 Years**	**p for Trend**
Other cardiovascular disease					
No. of deaths	402	422	153	121	
Age-adjusted HR (95% CI)	1.00 (ref)	1.02 (0.89, 1.16)	1.28 (1.06, 1.54)	1.36 (1.11, 1.67)	**<0.001**
Multivariable HR (95% CI)[a]	1.00 (ref)	1.05 (0.91, 1.20)	1.18 (0.97, 1.42)	1.17 (0.95, 1.44)	0.06
Multivariable HR (95% CI)[b]	1.00 (ref)	1.04 (0.90, 1.19)	1.14 (0.94, 1.38)	1.13 (0.92, 1.39)	0.14

Note: Boldface indicates statistical significance (p ≤ 0.05).

[a] Multivariable model adjusted for age (continuous), alcohol consumption (none, 0.1–4.9, 5.0–14.9, ≥15.0 g/day), physical exercise (metabolic equivalent values; quintiles), multivitamin use (yes, no), menopausal status (premenopausal, postmenopausal) and post-menopausal hormone use (never, past, and current), physical exam in the past 2 years (no, yes for symptoms, and yes for screenings), healthy eating score (quintiles), smoking status (never, former, and current), pack-years (<10, 10–19, 20–39, ≥40 for former smokers; <25, 25–44, 45–64, ≥65 for current smokers), BMI (<21, 21–22.9, 23–24.9, 25–27.4, 27.5–29.9, 30–34.9 ≥35), and husband's education (less than high school, some high school, high school graduate, college, graduate school, missingness).

[b] This model was further adjusted for hypertension (yes, no), hypercholesterolemia (yes, no), and type 2 diabetes (yes, no). HR, hazard ratio; NHS, Nurses' Health Study; CVD, cardiovascular diseases.

TABLE 1.8 Association between Night-Shift Work and Selected Endpoints Stratified by Smoking Status, 1988–2010[a]

Mortality	Night-Shift Work Duration					
	Never	1–5 Years	6–14 Years	≥15 Years	p-Value for Trend	p-Value for Interaction[b]
All causes						
Never smokers						
No. of deaths	1,863	1,833	638	457		
HR (95% CI)	1.00 (ref)	1.04 (0.98, 1.11)	1.20 (1.10, 1.31)	1.04 (0.94, 1.16)	**0.05**	
Former smokers						
No. of deaths	2,724	2,728	916	657		
HR (95% CI)	1.00 (ref)	0.98 (0.93, 1.03)	1.06 (0.98, 1.14)	1.09 (1.00, 1.19)	**0.01**	0.15
Current smokers						
No. of deaths	670	734	293	243		
HR (95% CI)	1.00 (ref)	1.07 (0.96,1.19)	1.08 (0.94, 1.24)	1.21 (1.04, 1.40)	**0.02**	
All cardiovascular diseases						
Never smokers						
No. of deaths	411	430	154	139		
HR (95% CI)	1.00 (ref)	1.11 (0.97, 1.27)	1.25 (1.04, 1.51)	1.25 (1.03, 1.52)	**0.01**	
Former smokers						
No. of deaths	516	521	199	146		
HR (95% CI)	1.00 (ref)	1.00 (0.89, 1.13)	1.17 (0.99, 1.38)	1.16 (0.97, 1.40)	**0.03**	0.40

(Continued)

TABLE 1.8 (Continued) Association between Night-Shift Work and Selected Endpoints Stratified by Smoking Status, 1988–2010[a]

		Night-Shift Work Duration				
Mortality	Never	1–5 Years	6–14 Years	≥15 Years	p-Value for Trend	p-Value for Interaction[b]
Current smokers						
No. of deaths	169	145	76	57		
HR (95% CI)	1.00 (ref)	0.86 (0.69, 1.08)	1.10 (0.83, 1.44)	1.11 (0.82, 1.50)	0.24	
Ischemic heart disease						
Never smokers						
No. of deaths	138	143	53	55		
HR (95% CI)	1.00 (ref)	1.11 (0.88, 1.40)	1.24 (0.90, 1.71)	1.31 (0.96, 1.81)	0.08	
Former smokers						
No. of deaths	192	175	71	62		
HR (95% CI)	1.00 (ref)	0.91 (0.74, 1.12)	1.08 (0.82, 1.42)	1.24 (0.93, 1.16)	0.08	0.26
Current smokers						
No. of deaths	63	57	41	28		
HR (95% CI)	1.00 (ref)	0.91 (0.63, 1.31)	1.63 (1.09, 2.43)	1.41 (0.90, 2.22)	**0.02**	
All cancer						
Never smokers						
No. of deaths	668	672	205	1,357		
HR (95% CI)	1.00 (ref)	1.05 (0.95, 1.17)	1.10 (0.94, 1.28)	0.94 (0.7, 1.13)	0.77	

(Continued)

TABLE 1.8 (Continued)   Association between Night-Shift Work and Selected Endpoints Stratified by Smoking Status, 1988–2010[a]

Mortality	Never	1–5 Years	6–14 Years	≥15 Years	p-Value for Trend	p-Value for Interaction[b]
			Night-Shift Work Duration			
Former smokers						
No. of deaths	1,085	1,102	339	235		
HR (95% CI)	1.00 (ref)	0.99 (0.91, 1.08)	1.01 (0.90, 1.15)	1.04 (0.90, 1.20)	0.58	0.18
Current smokers						
No. of deaths	300	342	117	118		
HR (95% CI)	1.00 (ref)	1.11 (0.94, 1.29)	1.00 (0.80, 1.24)	1.35 (1.09, 1.68)	**0.03**	
Lung cancer						
Never smokers						
No. of deaths	36	47	12	9		
HR (95% CI)	1.00 (ref)	1.31 (0.84, 2.03)	1.27 (0.65, 2.48)	1.09 (0.51,2.33)	0.80	
Former smokers						
No. of deaths	314	309	90	66		
HR (95% CI)	1.00 (ref)	0.99 (0.84, 1.16)	0.91 (0.71, 1.15)	0.96 (0.73,1.26)	0.57	0.17
Current smokers						
No. of deaths	140	149	60	67		
HR (95% CI)	1.00 (ref)	1.11 (0.86, 1.42)	1.16 (0.84, 1.61)	1.88 (1.36, 2.62)	**<0.001**	

(Continued)

TABLE 1.8 (Continued)   Association between Night-Shift Work and Selected Endpoints Stratified by Smoking Status, 1988–2010[a]

Mortality	Night-Shift Work Duration				p-Value for Trend	p-Value for Interaction[b]
	Never	1–5 Years	6–14 Years	≥15 Years		
Colorectal cancer						
Never smokers						
No. of deaths	64	63	23	20		
HR (95% CI)	1.00 (ref)	1.08 (0.76, 1.53)	1.30 (0.80, 2.12)	1.43 (0.85, 2.39)	0.13	
Former smokers						
No. of deaths	93	92	25	24		
HR (95% CI)	1.00 (ref)	0.98 (0.73, 1.32)	0.88 (0.56, 1.38)	1.19 (0.75, 1.89)	0.60	0.95
Current smokers						
No. of deaths	20	14	7	6		
HR (95% CI)	1.00 (ref)	0.74 (0.33, 1.67)	1.07 (0.40 2.87)	0.94 (0.32, 2.80)	0.89	

Note: Boldface indicates statistical significance (≤ 0.05).

[a] Cox proportional hazard model adjusted for age (continuous), alcohol consumption (none, 0.1–4.9, 5.0–14.9, ≥ 15.0 g/day), physical exercise (MET values; quintiles), multivitamin use (yes, no), menopausal status (premenopausal, postmenopausal) and postmenopausal hormone use (never, past, and current), physical exam in the past 2 years (no, yes for symptoms, and yes for screenings), healthy eating score (quintiles), smoking status (never, former, current), pack-years (<10, 10–19, 20–39 $\geq$40 for former smokers; <25, 25–44, 45–64, $\geq$ 65 for current smokers), BMI (<21, 21–22.9, 23–24.9, 25–27.4, 27.5–29.9, 30–34.9 $\geq$35), and husband's education (less than high school, some high school, high school graduate, college, graduate school, missingness).

[b] p-values for interaction were obtained by adding interaction terms into the models and performing likelihood ratio tests. HR, hazard ratio; NHS: Nurses' Health Study.

and a recent meta-analysis[55] are in line with the present findings. One cohort study of 504 papermill workers observed a significant increase in IHD risk after 10 years of shift work.[84] The other study of the NHS cohort, but with only 4 years of follow-up and many fewer cases (n = 292), observed an increase in coronary heart disease (both fatal and nonfatal) incidence among night-shift workers after both <5 and ≥5 years of rotating night-shift work.[101] Moreover, a recent meta-analysis[55] suggested that shift work increases the risk of coronary events (including IHD and other CVD, mixed incidence and mortality data); the association remained when restricted to higher-quality studies and was stronger for night-shift work, which disrupts circadian rhythms more than other types of shift work. Of the five studies accounting for duration of shift work,[71,101–104] three reported increased risks after longer durations of shift work,[71,101,102] with the exception of two case-control studies.[103,104] Together with previous evidence,[61,63] the current study provides further support for a link between night-shift work, especially of longer durations, and increased IHD events.

Several underlying biological mechanisms make such an association plausible: autonomic nervous system activation, an increased inflammatory state, changes in lipid and glucose metabolism, and resulting changes in the risk for atherosclerosis, all of which have previously been described in night-shift workers.[105] The finding that women with longer durations of rotating night-shift work had higher BMIs and tended to gain more weight since age 18 years may be intermediate conditions subsequent to night-shift work, and adjustment for them in multivariable models may underestimate associations. However, analyses stratified by BMI did not alter the results.

The moderately increased lung and colorectal cancer mortality among women with ≥15 years of night-shift work is consistent with previous reports on the risk of developing incident lung and colorectal cancer among night shift–working women in the same cohort.[106,107] Relatively few studies have examined associations between shift work and the risk of these two cancers.

For lung cancer, two studies[108,109] reported a higher risk among women in nursing, an occupation with a high prevalence of night-shift workers, even after adjustment for smoking. A less reliable study[68] found no association between night work and lung cancer risk, but this study was based on registry occupational data, hence possibly underestimating associations, and was not able to account for important confounders or effect modifiers, including smoking. Similarly, the same study found no association between night work and colorectal cancer. Among women with ≥30 years of rotating night-shift work, breast cancer mortality was nonsignificantly increased by 47% in the current study, which is somewhat consistent with a previous report in the same cohort[67] describing a significant 36% increase in breast cancer incidence. Current results also suggest potentially increased mortality from renal cancer and myeloma among women with >5 years of rotating night-shift work. The number of deaths was <100 for each of these cancers; therefore, power was likely limited in these analyses, warranting additional studies with larger sample sizes.

This study has a few limitations of note. First, night-shift work information was not updated after 1988, potentially leading to exposure misclassification because nurses may have changed their night-shift work status during the 22-year follow-up. However, much of follow-up was accrued at midlife or around retirement of these nurses, and more detailed job histories in the NHS2 cohort suggest that the percentage of nurses working rotating night shifts declines from roughly 40% in their early 20s to <5% after age 45 years, with only very few women (<2%) starting night shifts midlife or later (unpublished observation). A further limitation of the exposure assessment is that the timing of exposure in earlier adult life is not known (along with the time between exposure and disease diagnosis). Moreover, in the NHS2 cohort, where night-shift work history was assessed in much greater detail, <10% of all nurses worked permanent night shifts. A potential concern is that these nurses may not have classified themselves as working rotating night shifts in the current analysis, yet exclusion of these women did not change results of a prior analysis.[66] Nonetheless, any remaining nondifferential misclassification of the exposure assessment could have biased results, albeit likely toward the null in most situations. Another concern relates to multiple comparisons; however, when requiring a p-value of 0.01 for significance, the results remain largely unchanged. Last, it is difficult to differentiate the effect of night-shift work from lack of sleep; however, sensitivity analyses restricted to women who reported 6–8 sleep hours (not including nap times) per day did not change results materially (data not shown).

This study also has considerable strengths. It is one of the largest prospective cohort studies worldwide with a high proportion of rotating night-shift workers and long follow-up time. A single occupation (nursing) provides more internal validity than a range of different occupational groups, where the association between shift work and disease outcomes could be confounded by occupational differences. Shift work is more common in people of lower Socioeconomics status (SES), a risk factor for multiple diseases. Using a cohort of nurses with relatively homogeneous SES minimized potential confounding. Furthermore, husbands' education was adjusted to account for residual confounding by SES. Although self-reported duration of rotating night shifts cannot be validated, it is likely that results are accurate, because other self-reports were highly accurate in a similar cohort,[110] and previous validations of similar questions (e.g., electric blanket use)[111] have shown reasonable reproducibility. Moreover, the prospective study design eliminates recall bias. Finally, information on a range of lifestyle factors was prospectively and repeatedly collected and used for multivariable adjustment to minimize potential confounding by these factors.

In summary, this study confirmed associations of rotating night-shift work and total and CVD (particularly IHD) mortality, and suggested associations of shift work with multiple cancer sites.

These results add to prior evidence of a potentially detrimental effect of rotating night-shift work on health and longevity.

Lighting Schemes for Treating Circadian Sleep Disorders

Sleep disturbances in older adults are quite common. Seniors living in assisted living facilities are perhaps the best example population at risk of circadian disorders that could be alleviated by light treatment; due to age-dependent reduced retinal light exposures and fixed lighting conditions in their living environments, seniors are less likely to experience the necessary, robust 24-hour, light/dark pattern needed for circadian entrainment. Prescribed 24-hour light/dark patterns have been shown to entrain circadian rhythms and thereby alleviate some sleep and agitation issues common among seniors, including those with AD. A 24-hour lighting scheme that delivers high circadian stimulation during the daytime hours and low circadian stimulation in the evening hours and night-lights that provide perceptual cues to decrease falls risks at night has been proposed.[111] Exposure to bright white light (at least 2500 lx and as high as 8000 lx at the cornea) for at least 1 hour in the morning for a period of at least 2 weeks was found to improve or consolidate nighttime sleep of AD patients. Greater sleep efficiency at night decreased the need to sleep during daytime hours and, in some cases, reduced agitated behavior such as pacing, aggressiveness, and speaking loudly.[112–116] Moreover, uncontrolled exposure to bright white light (average of 1136 lx at the cornea) during the entire day improved rest/activity of AD patients.[117] Finally, evening exposure to 30 lx of blue light from LEDs peaking at 470 nm at the eye for 2 hours consolidated rest/activity rhythms and increased sleep efficiency of older people with and without AD.[118,119] It is important to note, however, that a recent study failed to show significant improvement in nighttime sleep or daytime wake of AD patients after exposure to 1 hour of morning bright white light (>2500 lx at the eye).[120] Although seniors are more likely to benefit from light exposures, the application of light needs to be such that they are getting the correct dose at their corneas as well as getting the light at appropriate times. As discussed below, tuning the spectrum of light to optimally stimulate the circadian system will allow for use of lower light levels, reducing the likelihood of glare and increasing the likelihood of compliance.

Jet-lag is a temporary desynchronization between master clock time and environmental time (light/dark). The symptoms include insomnia and/or hypersomnia, fatigue, poor performance, and gastrointestinal problems. Eastward travel generally results in difficulty falling asleep, and westward travel results in difficulty maintaining sleep. Adaptation to a new time zone is usually slower after eastward travel than after westward travel. There are two main reasons for that: (1) those traveling east need to advance their biological clock to readjust to the local time at their destination; the time that daylight is available at the final destination will promote phase delay of the master clock, and (2) it is easier for the timing of the biological clock to be delayed than advanced. One study[121] showed that a combination of advancing sleep schedules for 1 hour per day plus morning light treatment (½ hour of 5000 lx + half-hour of less than 60 lx at the cornea) for 3.5 hours advanced the phase of the biological clock by 1.5 to 1.9 hours in 3 days. Although the principles for applying light treatment for reducing jet-lag symptoms are known, the implementation of the light treatment may be a challenge. Airlines are starting to use colored light inside airplanes to improve mood, but it is probably very difficult to shift the circadian clock while one is inside the plane. *Because the circadian clock is slow to shift, users need to start treatment a few days before they are scheduled to travel.* However, because travelers have busy schedules, the likelihood of compliance is low. A personal light treatment device could be developed to increase the likelihood of compliance.

Delayed sleep phase disorder (DSPD) is a disorder of the timing of sleep; people suffering from DSPD typically go to bed late and wake up late (3–6 hours later than normal sleeping hours). This pattern interferes with people's normal functioning because they have difficulty waking up in the morning for work, school, and social obligations, and since they go to bed late, they do not sleep for as many hours as those going to bed at more normal hours. DSPD in adolescents is common and probably associated with hormonal changes that occur at puberty. The exact causes of DSPD are not actually known, but light exposure after minimum core body temperature and dim light during the evening have been shown to advance the phase of the master clock of persons with DSPD.[122] Using this information, two field studies[123,124] were very recently conducted to investigate the impact of light exposures on dim light melatonin onset (DLMO), a primary marker for the timing of the master clock, and on sleep duration for two populations of eighth graders. It was hypothesized for one study conducted in North Carolina that the lack of morning short-wavelength light (by wearing orange goggles) would delay the timing of the students' master clocks. For the other study, conducted in New York, it was hypothesized that exposure to more evening light in spring relative to winter would also delay the master clocks of adolescents. In both studies, as expected, the students exhibited delayed DLMO as a result of removing short-wavelength morning light and as a result of seasonal changes in evening daylight. Also as expected, both sets of adolescents exhibited shorter sleep times; because of the delay in the timing of the master clock, they fell asleep later but still had to get up at a fixed time in the morning. These two field studies clearly demonstrate that by controlling circadian light exposures, it is possible to practically and effectively control circadian time and thereby affect meaningful outcome measures like sleep duration.

Advanced sleep phase syndrome (ASPD) is another disorder of the timing of sleep, but unlike sufferers of DSPD, people with ASPD go to bed early and wake up early (3 to 6 hours earlier than normal sleeping hours). The cause of ASPD is still unknown, but it seems to be caused by genetic factors that result in a biological clock that runs with a period slightly shorter than 24 hours, instead of slightly greater than 24 hours, which is more normal.

ASPD is a much less common disorder than DSPD. As with DSPD, using light as a treatment for this syndrome can help regulate and promote circadian entrainment. Light exposure in the early evening and reduced light during the morning have been shown to phase delay minimum core body temperature.[125] It should be made clear that with the new WiFi, smart meters, and EMF-generating devices that now pollute the environment, many chemically and EMF sensitive patients have been found to have this syndrome. It is usually very detrimental to their improvement and state of recovery.

Seasonal affective disorder (SAD) is a subtype of depression, with episodes occurring during winter months and remitting during summer months.[126] Symptoms of SAD include depression, hypersomnia, weight gain due to increased carbohydrate cravings, social withdrawal, and even suicidal thoughts. It is believed that because daylight availability decreases in the winter at high latitudes, the number of people experiencing SAD increases as the latitude increases, affecting as much as 28% of the population living in places like Alaska.[126] "The winter blues" is a subtype of SAD and is even more common than SAD. The mechanisms of SAD are still unknown, and there are several competing hypotheses as to what causes SAD and how light can be used as a treatment. One of the competing hypotheses is that the late daybreak during winter months delays the circadian rhythms of those more susceptible to SAD; in this case, morning light is believed to be effective in treating symptoms of SAD. Another hypothesis is that the overall melatonin production of those suffering with SAD is greater during winter months than during summer months, which extends the amount of time during the 24-hour day that their bodies think it is nighttime; in this case, light in the early morning or evening is recommended. If a person is formally diagnosed with SAD by a general practitioner, insurance companies may pay for the cost of light treatment devices. A recent study[127] showed that 400 lx at the eye of blue light (λmax = 470 nm) was able to significantly improve SAD symptoms compared to red light, which was used for placebo control. It has been suggested, however, that the positive impact of light on SAD symptoms is simply a result of placebo effects. Although research results may seem contradictory regarding its effectiveness, light treatment is still the most common nonpharmacological treatment prescribed for SAD.[128,129]

Lighting Devices for Treating Circadian Sleep Disorders

Light treatment has been successfully used to reduce symptoms of the circadian sleep disorders described above. Light boxes using fluorescent light sources are the most common treatment devices available on the market. In general, light boxes provide a high amount of white light (2500–10,000 lx at the cornea). The main disadvantages of such devices include the necessity to remain in one place for the duration of the treatment and the fact that some users experience discomfort from having to stare into the bright light. Until recently, light treatment devices used "full-spectrum" light. More recently, products have become available that use narrow-band blue (λmax = 470 nm) and green (λmax = 500 nm) LEDs. Because the circadian system is maximally sensitive to short wavelengths, much lower levels of blue and green light can be effective. The small, versatile nature of LEDs has also facilitated the development of personal light-treatment devices, such as LED light goggles. Treatment devices such as these goggles can reduce the need to sit in front of a light box for extended periods of time and may lead to higher compliance in using light as a treatment option.

If compliance with light treatment devices is an issue, a dual ambient lighting scheme that maximizes circadian stimulation during the day (bright, bluish-white light) and minimizes it at night (dim, yellowish-white light) while maintaining good visibility at any time, can be used. This approach has been proposed for older adults' residences.[52] The ability to control the light level and spectrum is very important when it comes to providing good circadian lighting without compromising the visual requirements for seniors.

Dawn simulators are marketed primarily for people with seasonal depression; these devices are programmed to gradually increase the light levels in the morning hours, simulating the sunrise. While the evidence of effectiveness is limited, and light quantities achieved are often below those from other light treatment devices, dawn simulators may have therapeutic value for certain sleep disorders and might serve as a supplement to other light-treatment strategies. To accurately assess the efficacy for any light-treatment method, however, it is necessary to measure circadian light exposure throughout waking hours.

A low-cost alternative to commercial products would be for people to spend 20–30 minutes outdoors daily, preferably at about the same time each morning. Even an overcast day will produce 2500 lx at the cornea, and bright sunshine can produce exposures greater than 10,000 lx at the cornea.

According to Rea, M.S. and M.G Figueiro,[13] it should be noted that removing circadian light at certain times is just as important as being exposed to circadian light at other times; therefore, the use of personal light treatment devices that provide and remove short-wavelength light based on each individual's desired entrainment schedule is envisioned. The Daysimeter,[122] a circadian and photopic light meter, can be used together with a feedback control system that will let the user know when to receive and remove circadian light to maintain or adjust entrainment. Currently, all light-treatment devices and prescriptions are effectively "open-loop" control systems, which means that light is given at a prescribed clock time (i.e., morning or evening) without respect to the person's own internal clock or unexpected light exposures. An exciting prospect for personal lighting control systems to increase the effectiveness of light treatment will be the development of "closed-loop" feedback systems that ensure that light and dark are being delivered according to an individual's circadian time in the context of unexpected light exposures experienced in normal life activities.

Beds with natural wood slats and cotton layers with sponges and foam can also be used. These beds also have EMF screens that can ground the EMF emphasis, which allows people to sleep.

Solar Activity at Birth Predicted Infant Survival and Women's Fertility in Historical Norway

According to Skjaervo et al.,[123] ultraviolet radiation (UVR) can suppress essential molecular and cellular mechanisms during early development in living organisms, and variations in solar activity during early development may thus influence their health and reproduction. Although the ultimate consequences of UVR on aquatic organisms in early life are well known, similar studies on terrestrial vertebrates, including humans, have remained limited. Using data on temporal variation in sunspot numbers and individual-based demographic data (N = 8662 births) from Norway between 1676 and 1878 while controlling for maternal effects, socioeconomic status, cohort, and ecology, Skjaervo et al.,[123] showed that solar activity (total solar irradiance) at birth decreased the probability of survival to adulthood for both men and women. On average, the lifespans of individuals born in a solar maximum period were 5.2 years shorter than those born in a solar minimum period. In addition, fertility and lifetime reproductive success (LRS) were reduced among low-status women born in years with high solar activity. The proximate explanation for the relationship between solar activity and infant mortality may be an effect of folate degradation during pregnancy caused by UVR. These results suggest that solar activity at birth may have consequences for human lifetime performance both within and between generations.

According to Skjaervo et al.,[123] environmental factors during the early development of an organism can have downstream effects on the phenotypic quality and reproductive performance of that organism.[124] Several long-term studies on a wide variety of species,[125] including humans,[126-128] have revealed that the environment an organism is exposed to early in life may influence adult life-history traits, such as survival, fertility, and lifetime reproductive success. Individuals may differ in their sensitivity to stressors during early development, which can be influenced by gender and life stage. First, it is generally accepted that males are more vulnerable to environmental stressors than females,[129] and second, such effects may vary at different life stages, with greater vulnerability appearing during early development.[129]

Exposure to high levels of ultraviolet radiation is one such type of environmental stressor that can affect later survival and reproductive performance.[130,131] Levels of UVR vary with solar activity,[132] latitude/altitude,[133] and photoperiod.[134] The detrimental effects of high UVR exposure during development are unclear but may act via multiple molecular (degradation of folate and DNA damage) and cellular (membrane damage) mechanisms[133,135–137] in the developing organism. Such effects may lead to detrimental consequences later in life.[138] However, organisms can exhibit differential defenses against UVR damage, including behavior (avoidance), accumulation of photo-protective compounds (pigmentation, e.g., melanin and carotenoids), cellular defense mechanisms (DNA-repair and antioxidants),[7,139] and specific genotypes.[134] Moreover, the presence of such UVR defense mechanisms is often costly and indicates that UVR represents a potential environmental factor in life-history evolution.[136] Indeed, a number of aquatic studies indicate that ambient UVR exposure during early development affects life-history traits later in life.[130,131]

REFERENCES

1. Rea, W. J. 1994. *Chemical Sensitivity. Sources of Total Body Load.* Vol. 2. Boca Raton, FL: Lewis, pp. 655.
2. Hildebrandt, M. L. 2015. *Diurnal Patterns of Lower Atmospheric Pollution in Two Urbanized Valleys.* Dept of Geography. Southern Illinois University Edwardsville. Edwardsville, IL.
3. Ellis, A. W., M. L. Hildebrandt, W. M. Thomas, and H. J. S. Fernando. 2000. Analysis of the climatic mechanisms contributing to the summertime transport of lower atmospheric ozone within metropolitan Phoenix, Arizona. *Climate Research.* 15: 13–31.
4. Hildebrandt, M. L. 2000. A climatological analysis of lower atmospheric ozone transport across Phoenix. *Arizona, Papers of the Applied Geography Conferences.* 23: 145–153.

5. Berman, N. S., D. L. Boyer, A. J. Brazel, S. W. Brazel, R. R. Chen, H. J. S. Fernando, and M. J. Fitch. 1995. Synoptic classification and physical model experiments to guide field studies in complex terrain. *J. Appl. Meteor.* 34: 719–730.

6. Ellis, A. W., M. L. Hildebrandt, and H. J. S. Fernando. 1999. Evidence of lower atmospheric ozone 'sloshing' in an urbanized valley. *Phys. Geogr.* 20: 520–536.

7. Martinez-Levasseur, L. M., M. A. Birch-Machin, A. Bowman, D. Gendron, E. Weatherhead, R. J. Knell, and K. Acevedo-Whitehouse. 2013. Whales use distinct strategies to counteract solar ultraviolet radiation. *Sci. Rep.* 3: 2386.

8. URBAIR, 1997. Urban air quality management strategy in Asia: Kathmandu Valley Report, World Bank Technical Paper No. 378, J. Shah and T. Nagpal (eds.), World Bank.

9. Scientific Expers, 2006. *Air Quality Guidelines: Global Update 2005: Particulate Matter, Ozone, Nitrogen Dioxide, and Sulfur Dioxide. Copenhagen.* Denmark: World Health Organization Europe.

10. Stern, D. P. 1994. The art of mapping the magnetosphere. *Journal of Geophysical* 99(A9): 17169–17198.

11. Imagine the Universe! Dictionary. NASA. Archived from the original on 4 Feb 2015.

12. Havas, M. 2006. Electromagnetic hypersensitivity: Biological effects of dirty electricity with emphasis on diabetes and multiple sclerosis. *Electromagn. Biol. Med.* 25: 259–268.

13. Rea, M. S., and M. G. Figueiro. 2011. What is "healthy lighting? *J. Hi. Spe. Ele. Syst.* 20(02): 321–342.

14. Moore, R. Y. 1997. Circadian rhythms: Basic neurobiology and clinical applications. *Annu. Rev. Med.* 48: 253–266.

15. Cohen, R., and N. Kronfeld-Schor. 2006. Individual variability and photic entrainment of circadian rhythms in golden spiny mice. *Physiol. Behav.* 87(3): 563–74, Epub 2006 Feb 7.

16. Bullough, J. D., M. G. Figueiro, and M. S. Rea. 2006. Of mice and women: Light as a circadian stimulus in breast cancer research. *Cancer Causes and Control.* 17(4): 375–383.

17. Figueiro, M. G., J. D. Bullough, R. H. Parsons, and M. S. Rea. 2004. Preliminary evidence for spectral opponency in the suppression of melatonin by light in humans. *Neuro Report.* 15(2): 313–316.

18. Bullough, J. D., M. G. Figueiro, B. P. Possidente, R. H. Parsons, and M. S. Rea. 2005. Additivity in murine circadian phototransduction. *Zoolog Sci.* 22: 223–227.

19. Berson, D. M., F. A. Dunn, and M. Takao. 2002. Phototransduction by retinal ganglion cells that set the circadian clock. *Science*. 295: 1070–1073.

20. Hattar, S., R. J. Lucas, N. Mrosovsky et al. 2003. Melanopsin and rod-cone photoreceptive systems account for all major accessory visual functions in mice. *Nature*. 424: 75–81.

21. Foster, R. G., and M. W. Hankins. 2002. Non-rod, non-cone photoreception in the vertebrates. *Prog. Retin. Eye Res.* 21(6): 507–527.

22. Rea, M. S., M. G. Figueiro, A. Bierman, and J. D. Bullough. 2010. Circadian light. *Journal of Circadian Rhythms*. 8: 2.

23. Figueiro, M. G., A. Bierman, and M. S. Rea. 2008. Retinal mechanisms determine the subadditive response to polychromatic light by the human circadian system. *Neurosci. Lett.* 438(2): 242–245, Epub 2008 Apr 20.

24. Rea, M. S., M. G. Figueiro, J. D. Bullough, and A. Bierman. 2005. A model of phototransduction by the human circadian system. *Brain Res. Rev.* 50(2): 213–228.

25. Glovinsky, P. Spielman, A. 2005. *The Insomnia Answer: A Personalized Program for Identifying and Overcoming the Three Types of Insomnia*. New York: Perigee Books.

26. Arendt, J. 1995. *Melatonin and the Mammalian Pineal Gland*. 1st ed. London: Chapman & Hall.

27. Commission International de l'Éclairage (CIE), 1978. *Light As a True Visual Quantity: Principles of Measurement. No 41*. Paris: Commission Internationale de l'Éclairage.

28. M. S. Rea, ed., 2000. *IESNA Lighting Handbook: Reference and Application*. 9th ed. New York: Illuminating Engineering Society of North America.

29. Weale, R. A. 1953. Colour vision in the peripheral retina. *Br. Med. Bull.* 9: 55–60.

30. Commission Internationale de l'Eclairage (CIE), 1990. *CIE 1988 2°Spectral Luminous Efficiency Function for Photopic Vision*. Vienna: Commission Internationale de l'Eclairage.

31. Commission Internationale de l'Eclairage CIE), 2005. *CIE 2005 10° Photopic Photometric Observer*. Vienna: Commission Internationale de l'Eclairage.

32. Rea, M. S. 2002. "Light—Much More Than Vision," (Keynote). Light and Human Health. *EPRI/LRO 5th International Lighting Research Symposium*, Palo Alto, CA: The Lighting Research Office of the Electric Power Research Institute, 1–15.

33. Rea, M. S., M. G. Figueiro, and J. D. Bullough. 2002. Circadian photobiology: An emerging framework for lighting practice and research. *Light. Res. Technol.* 34(3): 177–190.
34. Rea, M. S., and M. J. Ouellette. 1991. Relative visual performance: A basis for application. *Lighting Research and Technology.* 23: 135–144.
35. Rüger, M., M. C. Gordijn, D. G. Beersma, B. de Vries, and S. Daan. 2005. Nasal versus temporal illumination of the human retina: Effects on core body temperature, melatonin, and circadian phase. *J. Biol. Rhythms.* 20(1): 60–70.
36. Van Derlofske, J., A. Bierman, M. S. Rea, J. Ramanath, and J. D. Bullough. 2002. Design and optimization of retinal flux density meter. *Meas. Sci. Technol.* 13: 821.
37. Glickman, G., J. P. Hanifin, M. D. Rollag, J. Wang, H. Cooper, and G. C. Brainard. 2003. Inferior retinal light exposure is more effective than superior retinal exposure in suppressing melatonin in humans. *J. Biol. Rhythms.* 18(1): 71–79.
38. McIntyre, M., T. R. Norman, G. D. Burrows, and S. M. Armstrong. 1989. Human melatonin suppression by light is intensity dependent. *J. Pineal. Res.* 6(2): 149–156.
39. Tassi, P., N. Pellerin, M. Moessinger, A. Hoeft, and A. Muzet. 2000. Visual resolution in humans fluctuates over the 24 h period. *Chonobiol. Int.* 17: 187–195.
40. Jewett, M. E., D. W. Rimmer, J. F. Duffy, E. B. Klerman, R. E. Kronauer, and C. A. Czeisler. 1997. Human circadian pacemaker is sensitive to light throughout subjective day without evidence of transients. *Am. J. Physiol.* 273: R1800–R1809.
41. Khalsa, S. B., M. E. Jewett, C. Cajochen, and C. A. Czeisler, 2003. A phase response curve to single bright light pulses in human subjects. 549(Pt 3): 845–852. Epub 2003 April 25.
42. Lynch, H. J., M. H. Deng, and R. J. Wurtman. 1985. Indirect effects of light: Ecological and ethological considerations. *Ann. NY Acad. Sci.* 453: 231–241.
43. Hebert, M., S. K. Martin, C. Lee, and C. I. Eastman. 2002. The effects of prior light history on the suppression of melatonin by light in humans. *J. Pineal. Res.* 33(4): 198–203.
44. Roenneberg, T., A. Wirz-Justice, and M. Merrow. 2003. Life between clocks: Daily temporal patterns of human chronotypes. *J. Biol. Rhythms.* 18(1): 80–90.
45. Hagenauer, M. H., J. I. Perryman, T. M. Lee, and M. A. Carskadon. 2009. Adolescent changes in the homeostatic and circadian regulation of sleep. *Dev. Neurosci.* 31: 276–284.

46. Harper, D. G., E. G. Stopa, V. Kuo-Leblanc, A. C. McKee, K. Asayama, L. Volicer, N. Kowall, and A. Statlin. 2008. Dorsomedial SCN neuronal subpopulations subserve different functions in human dementia. *Brain.* 131(Pt 6), 1609–1617. Epub 2008 Mar 27.
47. Wu, Y. H., and D. F. Swaab. 2007. Disturbance and strategies for reactivation of the circadian rhythm system in aging and Alzheimer's disease. *Sleep Med.* 8(6): 623–636, Epub 2007 Mar 26.
48. United States Department of Labor, Bureau of Labor Statistics, July 1, 2005. Workers on Flexible and Shift Schedules in May 2004, press release. USDL 05–1198.
49. Lowden, A., C. Moreno, U. Holmbäck, M. Lennernäs, and P. Tucker. 2010. Eating and shift work—effects on habits, metabolism and performance. *Scand J Work Environ Health.* 36(2): 150–162, Epub ahead of print.
50. Bøggild, H., and A. Knutsson. 1999. Shift work, risk factors and cardiovascular disease. *Scand J Work Environ Health.* 25(2): 85–99.
51. Figueiro, M. G., M. S. Rea, P. Boyce, R. White, and K. Kolberg. 2001. The effects of bright light on day and night shift nurses' performance and well-being in the NICU. *Neonatal Intensive Care.* 14: 29–32.
52. Cajochen, C. 2007. Alerting effects of light. *Sleep Med. Rev.* 11(6): 453–464.
53. Straif, K., R. Baan, Y. Grosse, B. Secretan, F. El Ghissassi, V. Bouvard, A. Altieri, L. Benbrahim-Tallaa, and V. Cogliano. 2007. Carcinogenicity of shift-work, painting, and fire-fighting. *Lancet Oncol.* 8: 1065–1066.
54. Kivimaki, M., G. D. Batty, and C. Hublin. 2011. Shift work as a risk factor for future type 2 diabetes: Evidence, mechanisms, implications, and future research directions. *PLoS Med* 8(12): e1001138.
55. Vyas, M. V., A. X. Garg, A. V. Iansavichus et al. 2012. Shift work and vascular events: Systematic review and meta-analysis. *BMJ.* 345: e4800.
56. Hansen, J. 2001. Increased breast cancer risk among women who work predominantly at night. *Epidemiology* 12(1): 74–77.
57. Hansen, J., and C. F. Lassen. 2012. Nested case-control study of night shift work and breast cancer risk among women in the Danish military. *Occup. Environ. Med.* 69(8): 551–556.
58. Jia, Y., Y. Lu, K. Wu et al. 2013. Does night work increase the risk of breast cancer? A systematic review and meta-analysis of epidemiological studies. *Cancer Epidemiol.* 37(3): 197–206.

59. Lie, J. A., H. Kjuus, S. Zienolddiny et al. 2011. Night work and breast cancer risk among Norwegian nurses: Assessment by different exposure metrics. *Am. J. Epidemiol.* 173(11): 1272–1279.

60. Lie, J. A., J. Roessink, and K. Kjaerheim. 2006. Breast cancer and night work among Norwegian nurses. *Cancer Causes Control.* 17(1): 39–44.

61. Menegaux, F., T. Truong, A. Anger et al. 2013. Night work and breast cancer: A population-based case-control study in France (the CECILE study). *Int. J. Cancer.* 132(4): 924–931.

62. O'Leary, E. S., E. R. Schoenfeld, R. G. Stevens et al. 2006. Shift work, light at night, and breast cancer on Long Island, New York. *Am. J. Epidemiol.* 164(4): 358–366.

63. Pesch, B., V. Harth, S. Rabstein et al. 2010. Night work and breast cancer—results from the German GENICA study. *Scand J Work Environ Health.* 36(2): 134–141.

64. Pronk, A., B. T. Ji, X. O. Shu et al. 2010. Night shift work and breast cancer risk in a cohort of Chinese women. *Am. J. Epidemiol.* 171(9): 953–959.

65. Pukkala, E., J. I. Martinsen, E. Lynge et al. 2009. Occupation and cancer—follow-up of 15 million people in five Nordic countries. *Acta Oncol.* 48(5): 646–790.

66. Schernhammer, E. S., C. H. Kroenke, F. Laden, and S. E. Hankinson. 2006. Night work and risk of breast cancer. *Epidemiology.* 17(1): 108–111.

67. Schernhammer, E. S., F. Laden, F. E. Speizer et al. 2001. Rotating night shifts and risk of breast cancer in women participating in the nurses' health study. *J. Natl. Cancer Inst.* 93(20): 1563–1568.

68. Schwartzbaum, J., A. Ahlbom, and M. Feychting. 2007. Cohort study of cancer risk among male and female shift workers. *Scand J Work Environ Health.* 33(5): 336–343.

69. Tynes, T., M. Hannevik, A. Andersen et al. 1996. Incidence of breast cancer in Norwegian female radio and telegraph operators. *Cancer Causes Control.* 7(2): 197–204.

70. Villeneuve, S., J. Fevotte, A. Anger et al. 2011. Breast cancer risk by occupation and industry: Analysis of the CECILE study, a population-based case-control study in France. *Am. J. Ind. Med.* 54(7): 499–509.

71. Karlsson, B., L. Alfredsson, A. Knutsson et al. 2005. Total mortality and cause-specific mortality of Swedish shift- and dayworkers in the pulp and paper industry in 1952–2001. *Scand J Work Environ Health.* 31(1): 30–35.

72. Straif, K., R. Baan, Y. Grosse et al. 2007. Carcinogenicity of shift-work, painting, and fire-fighting. *Lancet Oncol.* 8(12): 1065–1066.

73. Lockley, S. W., D. J. Dijk, O. Kosti et al. 2008. Alertness, mood and performance rhythm disturbances associated with circadian sleep disorders in the blind. *J. Sleep Res.* 17(2): 207–216.

74. Myers, J. A., M. F. Haney, R. F. Griffiths et al. 2014. Fatigue in air medical clinicians undertaking high-acuity patient transports. *Prehosp. Emerg. Care.* 19(1): 36–43.

75. Pan, A., E. S. Schernhammer, Q. Sun, and F. B. Hu. 2011. Rotating night shift work and risk of type 2 diabetes: Two prospective cohort studies in women. *PLoS Med.* 8(12): e1001141.

76. Rajaratnam, S. M., and J. Arendt. 2001. Health in a 24-h society. *Lancet.* 358(9286): 999–1005.

77. Ragland, D. R., B. A. Greiner, N. Krause et al. 1995. Occupational and nonoccupational correlates of alcohol consumption in urban transit operators. *Prev. Med.* 24(6): 634–645.

78. Rajaratnam, S. M., L. K. Barger, S. W. Lockley et al. 2012. Sleep disorders, health, and safety in police officers. *JAMA.* 306(23): 2567–2578.

79. Landrigan, C. P., J. M. Rothschild, J. W. Cronin et al. 2004. Effect of reducing interns' work hours on serious medical errors in intensive care units. *N. Engl. J. Med.* 351(18): 1838–1848.

80. Wang, X. S., M. E. Armstrong, B. J. Cairns et al. 2011. Shift work and chronic disease: The epidemiological evidence. *Occup Med (Lond).* 61(2): 78–89.

81. Yong, M., M. Nasterlack, P. Messerer et al. 2014. A retrospective cohort study of shift work and risk of cancer-specific mortality in German male chemical workers. *Int. Arch. Occup. Environ. Health.* 87: 1755–183.

82. Akerstedt, T., G. Kecklund, and S. E. Johansson. 2004. Shift work and mortality. *Chronobiol. Int.* 21(6): 1055–1061.

83. Shives, B., and B. Riley. 2013. Towards evidence based emergency medicine: Best BETs from the Manchester Royal Infirmary. BET 2: Is shift work bad for you? *Emerg. Med. J.* 30(10): 859.

84. Boggild, H., P. Suadicani, H. O. Hein, and F. Gyntelberg. 1999. Shift work, social class, and ischaemic heart disease in middle aged and elderly men; a 22 year follow up in the Copenhagen Male Study. *Occup. Environ. Med.* 56(9): 640–645.

85. Taylor, P. J., and S. J. Pocock. 1972. Mortality of shift and day workers 1956–1968. *Br. J. Ind. Med.* 29(2): 201–207.

86. Lange, T., S. Dimitrov, and J. Born. 2010. Effects of sleep and circadian rhythm on the human immune system. *Ann. N Y Acad. Sci.* 1193: 48–59.

87. Haus, E. L., and M. H. Smolensky. 2013. Shift work and cancer risk: Potential mechanistic roles of circadian disruption, light at night, and sleep deprivation. *Sleep Med. Rev.* 17(4): 273–284.

88. Fu, L., and C. C. Lee. 2003. The circadian clock: Pacemaker and tumour suppressor. *Nat. Rev. Cancer.* 3(5): 350–361.

89. van Dam, R. M., T. Li, D. Spiegelman, O. H. Franco, and F. B. Hu. 2008. Combined impact of lifestyle factors on mortality: Prospective cohort study in US women. *BMJ.* 337: a1440.

90. Green, C. B., J. S. Takahashi, and J. Bass. 2008. The meter of metabolism. *Cell.* 134(5): 728–742.

91. K-Laflamme, A., L. Wu, and S. Foucart. J. de Champlain. 1998. Impaired basal sympathetic tone and alpha1-adrenergic responsiveness in association with the hypotensive effect of melatonin in spontaneously hypertensive rats. *Am. J. Hypertens.* 11(2): 219–229.

92. Anwar, M. M., A. R. Meki, and H. H. Rahma. 2001. Inhibitory effects of melatonin on vascular reactivity: Possible role of vasoactive mediators. *Comp. Biochem. Physiol. C Toxicol. Pharmacol.* 130(3): 357–367.

93. Lucas, R. J., S. N. Peirson, D. M. Berson et al. 2014. Measuring and using light in the melanopsin age. *Trends Neurosci.* 37(1): 1–9.

94. Honma, S., D. Ono, Y. Suzuki et al. 2012. Suprachiasmatic nucleus: Cellular clocks and networks. *Prog. Brain Res.* 199: 129–141.

95. Filipski, E., and F. Levi. 2009. Circadian disruption in experimental cancer Q4 processes. *Integr. Cancer Ther.* 8(4): 298–302.

96. Sigurdardottir, L. G., U. A. Valdimarsdottir, K. Fall et al. 2012. Circadian disruption, sleep loss, and prostate cancer risk: A systematic review of epidemiologic studies. *Cancer Epidemiol. Biomarkers Prev.* 21(7): 1002–1011.

97. Ramin, C., E. Devore, W. Wang et al. 2014. Night shift work at specific age ranges and chronic disease risk factors. *Occup. Environ. Med.* 72(2): 100–107.

98. Boggild, H. 2009. Settling the question—the next review on shift work and heart disease in 2019. *Scand J. Work Environ. Health.* 35(3): 157–161.

99. Hublin, C., M. Partinen, K. Koskenvuo et al. 2010. Shift-work and cardiovascular disease: A population-based 22-year follow-up study. *Eur. J. Epidemiol.* 25(5): 315–323.

100. Thomas, C., and C. Power. 2010. Shift work and risk factors for cardiovascular disease: A study at age 45 years in the 1958. British birth cohort. *Eur. J. Epidemiol.* 25(5): 305–314.

101. Kawachi, I., G. A. Colditz, M. J. Stampfer et al. 1995. Prospective study of shift work and risk of coronary heart disease in women. *Circulation.* 92(11): 3178–3182.

102. Knutsson, A., T. Akerstedt, B. G. Jonsson, and K. Orth-Gomer. 1986. Increased risk of ischaemic heart disease in shift workers. *Lancet.* 8498: 89–92.

103. Yadegarfar, G., and R. McNamee. 2008. Shift work, confounding and death from ischaemic heart disease. *Occup. Environ. Med.* 65(3): 158–163.

104. McNamee, R., K. Binks, S. Jones et al. 1996. Shiftwork and mortality from ischaemic heart disease. *Occup. Environ. Med.* 53(6): 367–373.

105. Puttonen, S., M. Harma, and C. Hublin. 2010. Shift work and cardiovascular disease—pathways from circadian stress to morbidity. *Scand J. Work Environ. Health.* 36(2): 96–108.

106. Schernhammer, E. S., F. Laden, F. E. Speizer et al. 2003. Night shift work and risk of colorectal cancer in the Nurses' Health Study. *J. Natl. Cancer Inst.* 95(11): 825–828. http://dx.doi.org/10.1093/jnci/95.11.825.

107. Schernhammer, E. S., D. Feskanich, G. Liang, and J. Han. 2013. Rotating night shift work and lung cancer risk among female nurses in the United States. *Am. J. Epidemiol.* 178(9): 1434–1441.

108. Robinson, C. F., P. A. Sullivan, J. Li, and J. T. Walker. 2011. Occupational lung cancer in US women, 1984–1998. *Am. J. Ind. Med.* 54(2): 102–117.

109. Corbin, M., D. McLean, A. Mannetje et al. 2011. Lung cancer and occupation: A New Zealand cancer registry-based case-control study. *Am. J. Ind. Med.* 54(2): 89–101.

110. Colditz, G. A., P. Martin, M. J. Stampfer et al. 1986. Validation of questionnaire information on risk factors and disease outcomes in a prospective cohort study of women. *Am. J. Epidemiol.* 123(5): 894–900.

111. Figueiro, M. G. 2008. A proposed 24 hour lighting scheme for older adults. *Lighting Research and Technology.* 40: 153–160.

112. Koyama, E., H. Matsubara, and T. Nakano. 1999. Bright light treatment for sleep-wake disturbances in aged individuals with dementia. *Psychiatry and Clin. Neurosci.* 53: 227–229.

113. Lyketosos, C. G., L. Lindell Veiel, A. Baker, and C. Steele. 1999. A. randomized, controlled trial of bright light therapy for agitated behaviors in dementia patients residing in long-term care. *Int. J. Geriatr. Psychiatry.* 14: 520–525.

114. Mishima, K., Y. Hishikawa, and M. Okawa. 1998. Randomized, dim light controlled, crossover test of morning bright light therapy for rest-activity rhythm disorders in patients with vascular dementia and dementia of Alzheimer's type. *Chronobiol. Int.* 15: 647–654.

115. Lovell, B. B., S. Ancoli-Israel, and R. Gevirtz. 1995. Effect of bright light treatment on agitated behavior n institutionalized elderly subjects. *Psychiatry Res.* 57: 7–12.

116. Mishima, M., M. Okawa, Y. Hishikawa, S. Hozumi, H. Hori, and K. Takahashi. 1994. Morning bright light therapy for sleep and behavior disorders in elderly patients with dementia. *Acta Psychiatrica Scandinavica.* 89: 1–7.

117. Van Someren, E. J. W., A. Kessler, M. Mirmirann, and D. F. Swaab. 1997. Indirect bright light improves circadian rest-activity rhythm disturbances in demented patients. *Biol. Psychiatry.* 41: 955–963.

118. Figueiro, M. G., G. Eggleston, M. S. Rea et al. 2002. Effects of light exposure on behavior patterns of Alzheimer's disease patients: A pilot study. *5th LRO Res. Symp.,* Orlando, FL.

119. M. G. Figueiro, and M. S. Rea. 2005. LEDs: Improving the Sleep Quality of Older Adults. *Proceedings of the CIE Midterm Meeting and International Lighting Congress,* Leon, Spain, (May 18–21, 2005).

120. Dowling, G. A., R. L. Burr, E. J. Someren, E. M. Hubbard, J. S. Luxenberg, J. Mastick, and B. A. Cooper. 2008. Melatonin and bright-light treatment for rest-activity disruption in institutionalized patients with Alzheimer's disease. *J. Am. Geriatr. Soc.* 56(2): 239–46.

121. Eastman, C. I., C. J. Gazda, H. J. Burgess, S. J. Crowley, and L. F. Fogg. 2005. Advancing circadian rhythms before eastward flight: A strategy to prevent or reduce jet lag. *Sleep.* 28(1): 33–44.

122. Bierman, A., T. R. Klein, and M. S. Rea. 2005. The Daysimeter: A device for measuring optical radiation as a stimulus for the human circadian system. *Measurement Science and Technology.* 16: 2292–2299.

123. Skjaervo, G. R., F. Fossoy, and E. Roskaft. 2015. Solar activity at birth predicted infant survival and women's fertility in historical Norway. *Proceedings of the Royal Society B: Biological Sciences.* 282(1801).

124. Lindstrom, J. 1999. Early development and fitness in birds and mammals. *Trends Ecol. Evol.* 14: 343–348.
125. Metcalfe, N. B., and P. Monaghan. 2001. Compensation for a bad start: Grow now, pay later? *Trends Ecol. Evol.* 16: 254–260.
126. Bateson, P., D. Barker, T. Clutton-Brock et al. 2004. Developmental plasticity and human health. *Nature.* 430: 419–421.
127. Barouki, R., P. D. Gluckman, P. Grandjean, M. Hanson, and J. J. Heindel. 2012. Developmental origins of noncommunicable disease: Implications for research and public health. *Environ. Health.* 11: 42.
128. Lummaa, V., and T. Clutton-Brock. 2002. Early development, survival and reproduction in humans. *Trends Ecol. Evol.* 17: 141–147.
129. Clutton-Brock, T. H., S. D. Albon, and F. E. Guinness. 1985. Parental investment and sex differences in juvenile mortality in birds and mammals. *Nature.* 313: 131–133.
130. Bancroft, B. A., N. J. Baker, and A. R. Blaustein. 2007. Effects of UVB radiation on marine and freshwater organisms: A synthesis through meta-analysis. *Ecol. Lett.* 10: 332–345.
131. Llabres, M., S. Agusti, M. Fernandez, A. Canepa, F. Maurin, F. Vidal, and C. M. Duarte. 2013. Impact of elevated UVB radiation on marine biota: A meta-analysis. *Glob. Ecol. Biogeogr.* 22: 131–144.
132. Davis, G. E., and W. E. Lowell. 2008. Peaks of solar cycles affect the gender ratio. *Med. Hypotheses.* 71: 829–838.
133. Diffey, B. L. 1991. Solar ultraviolet radiation effects on biological systems. *Phys. Med. Biol.* 36: 299–328.
134. Lucock, M. Z. Yates, C. Martin et al. 2014. Vitamin D, folate, and potential early lifecycle environmental origin of significant adult phenotypes. *Evol. Med. Pub. Health.* 2014: 69–91.
135. Halliday, G. A. 2005. Inflammation, gene mutation and photoimmunosuppression in response to UVR-induced oxidative damage contributes to photocarcinogenesis. *Mutat. Res. Fundam. Mol. Mech. Mutagen.* 571: 107–120.
136. Jablonski, N. G., and G. Chaplin. 2000. The evolution of human skin coloration. *J. Hum. Evol.* 39: 57–106.
137. Dahms, H. U., and J. S. Lee. 2010. UV radiation in marine ectotherms: Molecular effects and responses. *Aquat. Toxicol.* 97: 3–14.
138. Juckett, D. A., and B. Rosenberg. 1993. Correlation of human longevity oscillations with sunspot cycles. *Radiat. Res.* 133: 312–320.

139. Hessen, D. O. 2002. *UV Radiation and Arctic Ecosystems*. Berlin, Germany: Springer, 17.
140. Sulman, F. G. 1976. *Health, Weather and Climate* P. 89, Basel, Switzerland: S. Karger.
141. U.S. Census Bureau. 2000. *Environmental Health Perspectives* 123(12): 1627.

CHAPTER 2

The Electromagnetic Phenomena as Incitants

S OLAR WINDS GENERATE THE earth's magnetic fields, as do the forces between the North and South Pole (Schumann's forces). Because of that, we have electromagnetism on earth; also, the earth has a giant magnetic field.

Overall, we can't avoid electricity since the earth behaves like an enormous electric circuit. The atmosphere is actually a weak conductor, and if there were no sources of charge, its existing electric charge would diffuse away in about 10 minutes. However, there are some complications. There is a "cavity" defined by the surface of the earth and the inner edge of the ionosphere 55 kilometers up.[1] At any moment, the total charge residing in this cavity is 500,000 coulombs[1] (electrostatic interaction between electrically charged particles). There is a vertical current flow between the ground and the ionosphere of $1-3 \times 10^{-12}$ amperes per square meter.[1] The resistance of the atmosphere is 200 ohms.[1] The current through a conductor between two points is directly

TABLE 2.1   Definitions of Earth Cavity

Coulombs—Electric static interaction between charged particulates
Ohms—Resistance of atmosphere
Amperes—Vertical current flow strength of magnetic field
Volts—Electromagnetic force
Amperes—Strength of magnetic field

proportional to the potential differences across two points. The voltage potential is 200,000 volts.[1] There are about 1000 lightning storms at any given moment worldwide, which can disturb segments of the atmosphere. Each episode produces 0.5 to 1 ampere, and these collectively account for the measured current flow in the earth's electromagnetic cavity (Tables 2.1 and 2.2).[1]

The Schumann resonances are quasi-standing electromagnetic waves that exist in this cavity.[1] Like waves on a spring, they are not present all the time, but have to be "excited" to be observed. They are not caused by anything internal to the earth, its crust, or its core. They seem to be related to electrical activity in the

TABLE 2.2   Abbreviations of EMF frequencies

ATC	Air traffic control
BM	Ballistic missile
CMRS	Commercial mobile radio service
DAB	Digital audio broadcast—terrestrial
DARS	Digital audio radio service—satellite
DMSP	Defense meteorological satellite program
FSS	Fixed satellite service
GPS	Global positioning satellite
GWCS	General wireless communications service
IMT2000	Third-generation mobile telephony
LEO	Low earth orbit
MDS	Multipoint distribution system
MSS	Mobile satellite service
NLMCS	New land mobile communications service
PCS	Personal communications service
TT&C	Tracking telemetry and command

atmosphere, particularly during times of intense lightning activity. They occur at several frequencies between 6 and 50 cycles per second[1]; specifically 7.8, 14, 20, 26, 33, 39, and 45 hertz, with a daily variation of about ±0.5 hertz.[1] As long as the properties of the earth's electromagnetic cavity remain about the same, these frequencies remain the same. Presumably, there is some change due to the solar sunspot cycle as the earth's ionosphere changes in response to the 11-year cycle of solar activity.[1] Schumann resonances are most easily seen between 2000 and 2200 uT.

Given that the earth's atmosphere carries a charge, a current, and a voltage, it is not surprising to find such electromagnetic waves. The resonant properties of this terrestrial cavity were first predicted by the German physicist W. O. Schumann between 1952 and 1957[1] and first detected by Schumann and Konig in 1954.[1] The first spectral representation of this phenomenon was prepared by Balser and Wagner in 1960. Much of the research in the last 20 years has been conducted by the Department of the Navy, which investigates extremely low-frequency communication with submarines (Table 2.3).[1]

Electromagnetic fields are found everywhere there is electricity.[2] Concerns about them center on the potential for very strong low fields to cause health effects in people exposed to them for long periods of time or at high intensities for shorter periods. *Groups of people thought to be at risk include power industry workers and people living close to voltage power lines, as well as people with neurodegenerative disease like polio, multiple sclerosis, Parkinson's, amyotrophic lateral sclerosis (ALS), and so* on who develop chemical sensitivity.

Over the last 25 years, a considerable amount of scientific research has been directed at determining the level of hazard

TABLE 2.3   Electromagnetic Waves

Hertz—Frequencies in cycle/second, changes with the solar cycle
Voltage—Electromagnetic force
Frequencies—Cycles/second

posed by EMFs. This has included both epidemiological studies (patterns of disease in groups of people) and laboratory studies on animals and human volunteers. The results of scientific research on the effects of EMFs on health are equivocal: many studies have been conducted that have found no links between high levels of exposure or proximity to power lines and health effects,[3] while other studies have reported statistical links.[3] In the meantime, power utilities have adopted a policy of "prudent avoidance" when building new electrical facilities and/or building in areas where there are milking cows because there has been harm shown in them.[3]

However, regardless of the state of scientific evidence, there is a perception among some sections of the public that there are health risks associated with exposure to strong and even weak EMFs. Many people perceive weather changes, full moons, overcast days, and so on.

Electric and magnetic changes from the earth, moon, and sun can have profound effects on human function. These can generate EMF smog when they emanate dirty electricity. These changes can cause behavioral changes, as has been seen for roughly centuries, in the form of lunatics (people who react aberrantly in the full phase of the moon). These aberrations of human behavior were seen when the full moon occurred each month. Now it has been shown that many people, especially those who have short-term memory loss and trouble concentrating, react to the blue frequency of sunlight. These people have trouble sleeping, and when they wear yellow glasses to screen out blue light, they sleep well and put out melatonin, the antioxidant that is eight times more potent than other antioxidants as well as being the sleep hormone. These patients are usually chemically sensitive and probably EMF sensitive.

It is well known that *many individuals have problems with weather and positively charged ions*; that is, one to three days before rain, due to positive ion effects, some people become ill and develop migraines, fatigue, difficulty remembering, and depression. Others are not as sharp and function at less than their optimum. In addition, they have problems with solar flares, sun spots, and other natural electromagnetic problems, causing some people to malfunction. All of these conditions act as EMF smog (dirty electricity).

The earth's own electromagnetic properties with gravity and other potentials are variable. These variable properties occur in different areas and terrains of its surface, which subtly influences human function, especially in the chemically sensitive and EMF-sensitive patient. Electrical or acoustic (even subaudio) frequencies from approaching weather fronts or thunderstorms may become troublesome for the chemically sensitive and EMF-sensitive patient. The EMF sensitive will react to these. Eventually, they may have a hypersensitivity to sunlight, as we have seen in children. Some babies even sneeze violently when initially exposed to sunlight. Some EMF patients can't tolerate sunlight and may have to wear dark glasses or avoid the sun altogether.

Now, with the advent of many technological inventions, we have human-made apparati beamed throughout the earth, which often plays havoc with human function and well-being. This includes radio waves, telephone, telegraph, television, computers, cell phones, Wi-Fi, and smart meters, which then do their own part to disrupt the earth's electromagnetic atmosphere and stability of the individuals, to their detriment, by generating dirty electricity.

In particular, many people who live along proposed high-voltage power lines are convinced that there are dangers associated with them and therefore do not want them to be constructed on or near their properties.[3] People are also concerned about the perceived risk to farm and native animals living near the power lines, especially stud breeding stock.[3] For example, one Missouri rancher lost 250 cattle due to dirty electricity beamed to his ranch. He also became electrosensitive.[3] The body runs with electromagnetic frequencies; therefore, because of the coherence phenomena of the physical environment, animals, food, vegetation, chemicals, and the rest of the environment communicate electromagnetically. Individuals can communicate with each other, and when there is excess, or the wrong frequencies are connected, the patient will have adverse reactions. These adverse reactions can range from mild dysfunctions to even debilitating dysfunction. The clinician may never think of EMF as the trigger of the disability unless he/she has often been reminded of EMF as having a possibility of being the triggering agent.

ELECTROMAGNETIC FIELDS

There are two different types of fields produced by electrical equipment and appliances: electric fields and magnetic fields. An electric field is an invisible force that relates to voltage, or the pressure under which electricity is forced along wires. Electric fields are present in any appliance plugged into a power point that is switched on, regardless of whether the appliance is turned on or off. Electric fields are strongest close to their source, but their strength rapidly diminishes as distance away from the source increases. They are blocked by many common materials such as wood or metal. Electric fields are measured in volts per meter or kilovolts (kV) per meter.

A magnetic field is an invisible force that is produced by the flow of electricity (commonly known as the current). Unlike electric fields, magnetic fields are only present when the electricity is on and the current is flowing. The strength of the magnetic field depends on the size of the current, and it also decreases rapidly as distance away from the source increases. Magnetic fields are usually measured in milligauss, but are sometimes measured in gauss, teslas, and microteslas (10 milligauss equals 1 microtesla). Unlike electric fields, however, magnetic fields are highly penetrative and difficult to shield.

Some terminology must be defined in order to understand electrical sensitivity, electromagnetism, and other clinical entities. These include the electric current and electromagnetism. Electric current goes under the influence of radio frequency. The electrical part is measured in volts/meter. Electrical fields or electrical charges tend to accumulate on opposite cell sources (charges) to form induced dipoles whose orientation changes with oscillations of the field. A dipole–dipole attraction occurs in the process. The attractive forces between the dipoles then align in the direction of the applied electrical field and form chains of many cells or molecules. These chains are mostly single-stranded but can be multistranded as well. (See Chapter 1, EMF section for more information.)

The term *magnetic potential* can be used for either of two quantities in classical electromagnetism: the *magnetic vector potential,* (A, the vector potential) and the *magnetic scalar potential* (Ψ) Both quantities can be used in certain circumstances to calculate the magnetic field. The magnetic field is measured by any *flow current,* the *tesla. Magnetic fields as low as one microtesla (a millionth of a tesla) can produce biological effects.*[4]

The more frequently used *magnetic vector potential,* A, is defined such that the curl of A is the magnetic B field. Together with the electric potential, the magnetic vector potential can be used to specify the electric field, E, as well. Therefore, many equations of electromagnetism can be written either in terms of E and B or in terms of the magnetic vector potential and electric potential. In more advanced theories such as quantum mechanics, most equations use the potentials and not the E and B fields.

The magnetic scalar potential, Ψ, is sometimes used to specify the magnetic H-field in cases when there are no free currents, in a manner analogous to using the electric potential to determine the electric field in electrostatics. One important use of Ψ is to determine the magnetic field due to permanent magnets when their magnetization is known. With some care, the scalar potential can be extended to include free currents as well.

The hertz is the unit of frequency in the international system of units. The hertz is equivalent to cycles/second of a periodic phenomenon. One of the most common uses is the description of the sine wave used in radio and audio applications. One hertz is one cycle/second, and 1 kilohertz is 10^3 hertz. A megahertz is 10^6 hertz, while a gigahertz is 10^9 hertz or cycles/second. A tetrahertz is 10^{12} cycles/second.

There are different configurations in order to neutralize electromagnetic currents so they don't damage the individual. These are solenoids and toroids.

Like all other electrical equipment, transmission lines produce both electric and magnetic fields. With power lines, the strength of the electric field varies with the operating voltage of the line, and the strength of the magnetic field is related to the amps, or current, flowing in the line. Field strengths are also related to the height of the lines, their geometric arrangement, and the arrangement of the power in multicircuit lines.

The question of whether EMFs can detrimentally affect biological systems has been addressed by many scientists over the last 20 years, with studies ranging from *in vitro* laboratory experiments on single cells to epidemiological studies on large populations. There are many thousands of primary research papers published in peer-reviewed journals, meta-analyses of groups of similar studies, and over 70 comprehensive secondary reviews carried out worldwide by professional committees and panels. Many studies show that EMF can affect biological systems. The grossest is death by electrocution or electroshock to start a stopped heart. Lower volts are used to stimulate paralyzed muscles or nerves, which can cause electrophysiological stimulation and hypersensitivity of response.

Our double-blind studies under environmentally controlled conditions at the EHC-Dallas show EMF sensitivity.

ELECTROMAGNETIC FIELD SPECTRUM

Understanding the electromagnetic spectrum and frequencies is necessary for the clinician to understand human function because the difference between the living and dead organism is its electromagnetic charges. The clinician can often shock the heart back to life, which is an example of the living organism response to EMF fields. The body runs on electromagnetic frequencies, and the EMF coherence phenomena between molds, foods, chemicals, and physical phenomena that these substances have with each other and EMF is essential to understanding the communication phenomena. Therefore, humans can be very sensitive to EMF waves, and they can mildly to severely affect human health and functions either positively or negatively.

We know that foods, molds, and chemicals in various forms can affect humans both positively or negatively depending on the form and circumstances. This phenomenon has been shown repeatedly by the technique of intradermal titration and neutralization, where antigens are diluted in 1 to 5, 1 to 25, 1 to 125, 1 to 625, 1 to 3000, and so on. This is done in order to demonstrate the patient's hypersensitivity response by provocation and then to find a lower treatment dose of antigen (frequencies) for therapy.

The electromagnetic spectrum identifies the names given to the various regions of the spectrum, along with their corresponding frequency (in cycles per second or hertz) and wavelength (in meters) (Figure 2.1). The longest wavelengths and the slowest frequencies are referred to as extremely low frequency or ELF. Electricity at 50/60 hertz falls within this range, as does brain wave activity (4–30 Hz) and heart rate (1–2 Hz). The next band is radio frequency and includes all forms of broadcast frequencies used for radio and television (thousands and millions Hz). Microwaves are smaller waves at higher frequencies (millions and billions Hz). Within this range are mobile phones, microwave ovens, and radar. Above this region comes *infrared*, which we associate with heating. Some remote controls rely on infrared to turn TVs on or off and to change channels. Microwaves are used for some saunas, but care must be taken not to burn the flesh. However, some chemically sensitive and EMF-sensitive patients cannot tolerate them. *Visible light* is part of the electromagnetic spectrum, and it is the form of energy we can see. *Ultraviolet*

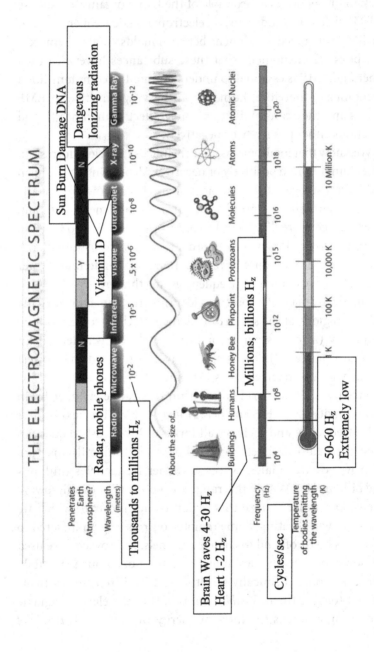

FIGURE 2.1   Electricity and magnetism in nature. These influence how the body works.

radiation is invisible to the human eye but can be "seen" by some species of insects. UV radiation can cause sunburn and damage DNA, which can lead to skin cancer, but also generates Vitamin D, which has many very important health benefits. Radiation at and above UV is *ionizing radiation*, which is very dangerous. For that reason, we limit human exposure to these frequencies and use them only when required for x-raying bone fractures or dental decay, for example.

Coherence

There are coherent links of EMF between the chemical, biological (mold, bacteria, virus, toxic chemicals, food, parasites), and physical EMF fields that allow connections and communications. This coherence is vital for EMF health (bone healing, heart rhythms, muscular-nerve conditions) and proper functions of intradermal neutralizations (EEG, EKG, EMG). There are, therefore, certain frequencies of EMF in individuals that can depend on adverse effects (dirty electricity) and the total body pollutant load of molds, mycotoxins, pollens, foods, chemicals, bacteria, and viruses that can be a determinant.

There is a federal use of the EMF spectrum. Table 2.4 shows how the EMF spectrum is divided by federal law. The non–federal controlled spectrum is about 30.4%, whereas the federal-controlled spectrum is 31.5%. There are shared bands of 39.9% that are below 6 gigahertz.

As one can see, the United States has made extreme use of the EMF spectrum in an orderly fashion. However, there is still much confusion.

Figure 2.2 shows the electromagnetic spectrum.

TABLE 2.4   EMF Spectrum

1. Non-federal controlled	30.4%
2. Federal controlled	31.15%
3. Shared bands	39.9%

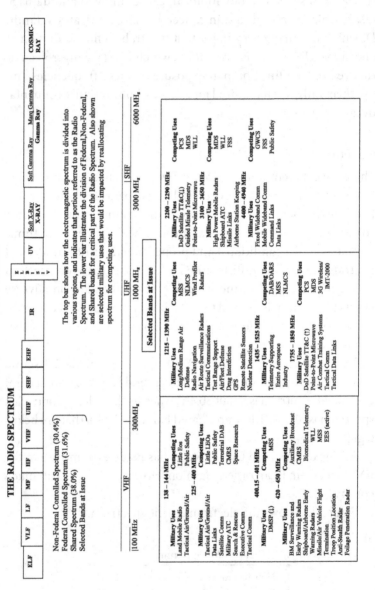

FIGURE 2.2   Shows the electromagnetic spectrum

ELECTROMAGNETIC FIELD WAVES

EM waves exist in a medium without matter: EMF waves have the following physical properties, (1) frequency, (2) wavelength, (3) photon energy. Each one is discussed separately.

Frequency of the Carrier Wave

An analog frequency occurs in the carrier frequency, inducting the center (normal) frequency or frequency of a carrier wave (radio waves). It is the nominal or center frequency of various kinds of radio signals with digital or physical modulation or double sideband carrier transmission (DOB/SG) of the AM-suppressed radio wave. The frequency of the unmodulated EM wave at the output of conventional amplitude-modulated (AM—unsuppressed carrier), frequency-modulated (FM), or phase-modulated (PM) radio transmitter may be present.

Radio Frequency

In radio frequencies, to receive radio signals, an antenna must be used. However, since the antenna will pick up thousands of radio signals at a time, a *radio tuner* is necessary to tune in to a particular frequency (or frequency range).[5] This is typically done via a resonator—in its simplest form, a circuit with a capacitor and an inductor, the property of conduction by which a change in current flow creates a voltage (electromagnetic force) in both the conductor and nearby conductors, forming a tuned circuit. The resonator amplifies oscillations within a particular frequency band while reducing oscillations at other frequencies outside the band. The tuner may isolate or magnify dirty electricity if that frequency contains any.

Often, tooth fillings, orthopedic or heart implants, metal prosthetics, and other metal implants can act as antennas, magnifying dirty electricity, making the patient ill. Often, mental haziness, brain fog, weakness and fatigue are the result environmental health center (EHC-Dallas).

Frequency modulation (FM) is the coding of information in a carrier wave where the frequency varies and the amplitude remains constant. This wave is compared with the *amplitude modulation* (AM), in which the amplitude of the carrier wave varies while the frequency remains consistent.

Frequencies range from 2.4×10^{23} Hz (1 Ge V gamma ray) down to the local plasma frequency of intrastellar median (r/KHz). *Audio frequency* (RF) is a rate of oscillation in the range of about 3 kHz to 300 GHz, which corresponds to the frequency of radio waves and the alternating currents that carry radio signals. RF usually refers to electrical rather than mechanical oscillations; however, mechanical RF systems do exist.

Although radio frequency is a rate of oscillation, the term *radio frequency* and its abbreviation RF are also used as synonyms for radio, that is, to describe the use of wireless communication, as opposed to communication via electric wires.

Special Properties of Radio Frequency Current
Electric currents that oscillate at radio frequencies have special properties not shared by direct or alternating currents of lower frequencies. The energy in an RF current can radiate off a conductor into space as electromagnetic waves (radio waves); this is the basis of radio technology. RF current does not penetrate deeply into electrical conductors but tends to flow along their surfaces; this is known as the *skin effect*. For this reason, when the human body comes in contact with high-power RF currents, it can cause superficial but serious burns called RF burns. RF currents applied to the body often do not cause the painful sensation of electric shock as do lower frequency currents.[6,7] This is because the current changes direction too quickly to trigger depolarization of nerve membranes.

EM radiation interacts with matter in different ways in different parts of the spectrum. The types of interactions can be sufficient that it seems justified to refer to different types of radiation. At the same time, there is a continuum containing all these different modes of EM radiation. *RF current can easily ionize air, creating a conductive path through it.* This property is exploited by "high-frequency" units used in electric arc welding, which use currents at higher frequencies than power distribution, which uses lower intensities, increasing muscle relaxation. Another property is the ability to appear to flow through paths that contain insulating material, like the dielectric insulator of a capacitor (the ability of the body to store an electrical charge).

When conducted by an ordinary electric cable, RF current has a tendency to reflect from discontinuities in the cable such as connectors and travel back down the cable toward the source, causing a condition called standing waves, so RF current must be carried by specialized types of cable called a *transmission line.*

Wavelength

Wavelengths propagate at the speed of light and have both electric and magnetic properties that are perpendicular to each other. The wavelength is inversely proportional to the wave frequency, so gamma waves have short wavelengths (fractions of atoms), whereas wavelengths at the other end of the spectrum can be as long as the universe, and when they contain dirty electricity, any length can cause dysfunction of metabolism.

The wavelength can be decreased; that is, radio diminishes when mechanically blocked by a mountain or building. Classification of the electromagnetic spectrum radiation is by wavelength into radio waves, microwaves, and (submillimeter) radiation, hertz, infrared, and the visible region, where we perceive light, and those we don't perceive as light: ultraviolet, x-rays, and gamma rays. When EM radiation interacts with a single atom or molecule, its behavior depends on the amount of energy per quantum (photon) it carries. The behavior of EM radiation depends on the wavelength.

Photon Energy

Energy photons are particles traveling through space carrying radiant energy.

EMR is emitted and absorbed by charged particles. *Photon energy* is directly proportional to wavelength frequency (gamma waves).

Photons have heightened energy (billion electron volts). Radio wave photons have very low energy (femto electron volts). When EM radiation interacts with a single atom or molecule, its behavior depends on the amount of energy per quantum (photon) it carries.

Therefore, depending on these three modalities, EMF, depending on the total body pollutant load and its response mechanisms, will *act as an access switch to cause clinical and basic science dysfunction.* But it can also be used to heal wounds and right dysfunctional physiology if the frequency is right for the particular individual being treated.

Electronics are responsible for the emission of most EMR because they have a low mass and are easily accelerated by a variety of mechanisms.

HISTORY OF THE ELECTROMAGNETIC SPECTRUM

Once, light was the only known part of the EM spectrum. The ancient Greeks studied light and its properties. It wasn't until scientific experiments almost several thousand years later that we discovered new findings about the EMF.

In 1800, Herschel[8] discovered infrared light, and the next year, Ritter[9] described invisible light rays that induced chemical variations. In 1845, Faraday[10] linked EM to the polarization of light traveling through a transparent material responding to an EM field. In the 1860s, Maxwell[11] developed a partial differential equation for the EM field. This observation led to the inference that light itself was an EM wave. His equation predicted an infinite number of frequencies of EM waves, all traveling at the speed of light. Hertz[12] built an apparatus to generate and detect radio waves. He also produced and measured the properties of the microwave. The knowledge of these new types of waves paved the way for the wireless telegraph and the radio. Rontgen[13] noticed x-rays when experimenting with high-voltage radiation in a evacuated tube. Villard[14] studied the radioactive omission of radiation-identified x and b particles with the power being greater than either (gamma rays). Audrade[15] measured the length of gamma rays and found they were shorter with higher frequencies than x-rays. Of course, today, we have a myriad of aberrative and technical changes in this field.

In 1980–1984, Smith and Monro[16] elaborated on and applied Frolich's concept of coherence for foods, molds, chemicals, and physical phenomena. This common EMF allowed communication between these different varieties. Rea[17] rapidly took this concept up, applying it to many patients. Rea[17] also did a double-blind study showing the existence of EMF changing and reproducing human function. Many researchers applied part of the medical spectrum to EMF- and chemically sensitive patients, like Schuman,[18] Hardell and Sage,[19] and Belpomme et al.[20] Several Eastern European scientists and clinicians looked into the idea of EMF causing or propagating disease. Electroshock for starting hearts, EMG, nerve stimulation for enhancing nerve and muscle function, and many other electric measurements occurred.

Figure 2.3 shows the health status of EMF patients from the beginning of the discovery of electricity until 2016. There is an

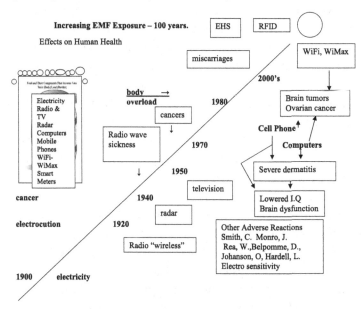

FIGURE 2.3   Increasing EMF exposure—100 years.

increase in sickness as the earth becomes more polluted with electrical equipment. Now, there is a crowding of wireless Wi-Fi towers across the United States and Europe and spreading to other continents. This wireless technology has not only caused electrosensitivity but also a large residence of neurovascular disease, depression, short-term memory loss, suicidal ideation, arteriosclerosis, and cancer.

ELECTROMAGNETIC INTERACTION SITES IN THE BODY WITH FEEDBACK

Nerve and EMF: The sensory nerves are the primary input parts that allow EMF to enter the body, being regulated by the ground regulation system. The autonomic nerves are part of this regulation system. Chemically and EMF-sensitive patients are examples of being sensitized through these routes. Heart shock for defibrillating a stopped heart is a good example of the necessity of proper EMF rhythm.

Sensory Intake: Anatomical parts are the skin; nervous system, especially brain; and the optic nerves, but the peripheral neurovascular system and total body sensors also constitute the ground regulation system. Autonomic nerves, especially those on the blood vessels and gastrointestinal (GI) organs, involve all of the involuntary nerves, allowing EMF to enter the body being regulated by the ground regulating system. These autonomic EMF waves are also triggered by food and water. They involve sympathetic and parasympathetic responses that are supposed to balance each other when triggered, but don't in the electrically and chemically sensitive patient.

Chemical Intake is primarily from olfactory odor sensitivity of the nose, but can also accumulate in the skin, GI tract, genitourinary (GU) tract, and respiratory areas (Figure 2.4).

The EMF sensory system is a global entity affecting every cell in the body, especially including the ground regulation systems (Figure 2.2). However, the brain and peripheral neurovascular and sensory systems appear to be most active. EMF affects many areas in the brain, including the olfactory apparatus, cerebral cortex, pineal gland, hypothalamus, amygdala, hippocampus, and the rest of the limbic system. The autonomic centers in the brain stem and the endocrine organs are included (Figures 2.5 and 2.6).

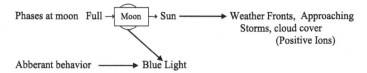

FIGURE 2.4   The ground regulation system skin peripheral sensory—olfactory → cerebral cortex.

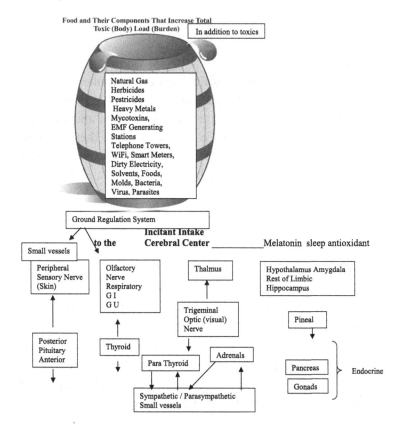

FIGURE 2.5   The ground regulation system intaking the total body load.

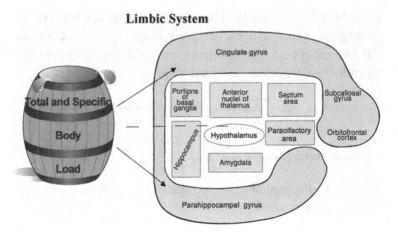

FIGURE 2.6   Limbic system

It needs to be emphasized again that there is a coherence effect between the chemical, physical, and biological systems using EMF waves in the body, and there are functional and corporal codes. There is an overlap of frequencies in all of these entities, allowing for positive healing frequency and negative effects of dirty electricity. Depending on which are triggered and dominant, they present as health or disease.

Total Body Pollutant Load

The total body pollutant load is made up of the sum total of pollutants that enter and are retained in the body, which has been emphasized in this series of books (Figure 2.4). Again, it has effects on electromagnetic sensitivity and long-term tissue changes. Some changes and symptoms occur immediately, while others, like degenerative neurovascular phenomena, arteriosclerosis, and cancer, occur after years of adverse pollutant exposure. As stated previously, some immediate changes occur with weather conditions in many people with chemical and EMF sensitivity.

TABLE 2.5   EMF Smog (Dirty Electricity) Changes on Earth
That Can Alter Human Function

(1)	Natural EMF changes emanating from the moon, sun spots, weather changes
(2)	Historical health of man with the induction of electricity
(3)	Effects of electricity on sensory system
(4)	EMF on the immune system
(5)	ELF and radio frequency
(6)	Stem cells and microwaves
(7)	Cell phones
(8)	Wi-Fi
(9)	Dairy cow perceiving ground currents
(10)	Lighting, grounding—protection of central system
(11)	Ground potential
(12)	Involvement of sensory system
(13)	Involvement of ANS and energy
(14)	Relation with acupuncture points
(15)	Smart meters
(16)	Computers

EMF from Earth, Moon, Sun, and Weather Patterns

Weather patterns have extremely important effects on the chemically and electrically sensitive individual. The sections on controlled weather conditions earlier in this volume and Chapter 1 provide an in-depth discussion of the responses of the chemically and electrically sensitive individual to variations in the weather. However, the following section emphasizes the basic science that one needs to understand in order to determine the effects of weather on the chemically and electrically sensitive (Table 2.5).

THE HUMAN-MADE ELECTRICAL ENVIRONMENT PRODUCING CHANGES IN HOMEOSTASIS

The outdoor and indoor human-made electrical environment can pose many problems for electrically sensitive individuals. For example, fluorescent lighting and lasers at checkouts may make shopping difficult, particularly if inhalants such as chemicals on in-store fabrics and objects provide an initial chemical sensitization. Thus, the EMF-sensitive individual is made more vulnerable to subtle electrical output. The patient may experience problems when near any electrical equipment such as electrical generating stations; substations; telephone towers; cell towers; high-tension power lines; radio, TV, or mobile phone transmitters; tape or DVD recorders; computers; mobile phones; Wi-Fi; or smart meters. This means that there can be many frequencies, each proportional to a velocity that the system will support. One such velocity is the velocity of light.

Another velocity is that with which the coherence propagates, on the order of meters per second. The fractal ratio is of the order of 10^8. One consequence of coherence is that with a coherent system, frequency becomes a fractal-like quantity with no single absolute value for its effects. This links the chemical, to the technological, to the biological frequency bands with respect to frequency effects. If there were not a duality between frequency and the chemical bond, spectroscopic analysis would be impossible.

According to Smith,[21] (1) it should be possible to erase the common frequency pattern of adaptation to exposure substances, including disease-generating and other triggers (toxic chemicals, biologics, EMF, medication), in an individual (shown by Tues in all such persons). (2) This would then leave the individuals' distinct frequency sensitivity patterns that could be measured and altered. These sensitivity patterns could be neutralized or erased on an individual basis, but this would only be a temporary measure as the underlying chemical and EMF sensitivities would reappear if not corrected by permanent tissue changes after total removal by cessation. (3) *Chemical detoxification by environmental medicine techniques would be necessary for a more permanent solution, which would include replacement of nutritional deficits.* However, some substances, such as metal or polysynthetics, especially implants such as polyurethane, polyvinyl, and polyethylene that can't be removed, need to have injections of the neutralizing dose every 4–7 days or more frequently to maintain stability due to the acquired hypersensitivity in the patient.

A protective measure for the EMF-sensitive individual would be for persons exposed to smart meters to avoid all foods and drinks containing aspartame or anything to which they are sensitive such as dental amalgams or other metals or synthetics. These can then be neutralized by frequencies imprinted into water and inserted into the patient every 1 to 4 days.[9] At times, this procedure is very successful, but at times it can be overridden. It would be equitable for the cost of such remedies to be born by the utilities involved as the price for making smart meter–exposed customers compatible with their equipment. Of course, this is never done. These metals also act as antennas—intercepting the output of the smart meters, radio frequencies, and many EMF antenna-prone devices and continuously triggering the body's adverse response. Plastics and other synthetics appear to disrupt EMF frequencies in an unknown way.

As shown previously, the electromagnetic spectrum is the range of all frequencies of EMF radiation. The EMF spectrum of an object is the characteristic distribution of EMF radiated or absorbed by the particular object. The EMF spectrum extends from very low frequencies (nanoteslas) to radio waves to gamma radiation and short wave (high frequencies). This allows the electrically sensitive individual to pick up many frequencies, resulting in adverse effects in the body. The limit of the long wavelengths is the size of the universe, while the other end of the spectrum is very short and at times also devastating (Figure 2.1).

Understanding this spectrum and frequencies and the coherence phenomena between physical phenomena; food; chemicals; and biologics such as molds, pollens, bacteria, parasites, and viruses is necessary to understand human function because of the difference between the living and dead organism is its EMF charge. Patients can develop responses and illness from inappropriately applied electrical impulses.[9]

Solenoids and toroids are different EMF configurations of the spectrum that neutralize EMR currents. Each will now be discussed.

Solenoids

Solenoids are pipe channels. They are a coil wound into a tightly packed helix, usually wrapped around a metallic core. The solenoid produces a uniform magnetic field in a volume of space when an electric current is passed through it. If the purpose of the solenoid is to impede changes in the electric current, it can be more specifically classified as an indicator rather than an electromagnet. Not all electromagnets are solenoids.

Transducers Convert Energy into Linear Motion
(See Figure 2.7.)

Magnetic fields are created by a seven-loop solenoid (cross-sectional view) described using field lines; this is why telephone cords are a multiloop cord: to cancel out the severe electric current running through them.

Toroidal Inductors and Transformers
In mathematics, a toroid is a doughnut-shaped object, such as an O-ring (Figure 2.8). It is a ring form of a solenoid. Its annular shape is generated by revolving a plane geometrical figure about an axis external to that figure that is parallel to the plane of the figure and does not intersect the figure (Figure 2.8). When a rectangle is

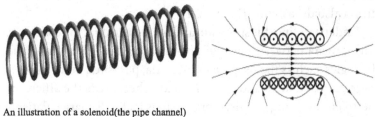

An illustration of a solenoid(the pipe channel)

FIGURE 2.7   An illustration of a solenoid (the pipe channel).

Toroid using a square

A torus is a type of donut

FIGURE 2.8   Toroid—doughnut-shaped object.

rotated around an axis parallel to one of its edges, then a hollow cylinder (resembling a piece of straight pipe) is produced. The toroid is a surface generated by rotating a closed curve about an axis that is in the same place as the curve that does not intersect it.

Toroidal inductors are the property of a conduction by which a change in current flow creates a voltage (electromagnetic force) in both the conductor and nearby conductors, and transformers, which are electronic components typically consisting of a circular ring-shaped magnetic core of iron powder, ferrite, or other material around which wire is coiled to make an inductor. Toroidal coils are used in a broad range of applications, such as high-frequency coils and transformers. Toroidal inductors can have higher Q factors and inductance than similarly constructed solenoid coils. This is due largely to the smaller number of turns required when the core provides a closed magnetic path.

HEALING PROPERTIES OF ELECTROMAGNETIC FIELDS

The magnetic flux in a high-permeability toroid is largely confined to the core; the confinement reduces the energy that can be absorbed by nearby objects, so toroidal cores offer some self-shielding. However, due to its permeability, it can accelerate the ion exchange of some tissues, even mitochondria, causing alteration of their function when exposed to some electromagnetic frequencies. Oftentimes, Ca^{++} influx into a permeable cell results in hypersensitivity, up to 1000 times normal. But these frequencies can also cause weakness and fatigue. However, at times when the physiology is right, these EMF frequencies can accelerate healing.

In the geometry of torus-shaped magnetic fields, the poloidal flux direction threads the "donut hole" in the center of the torus, while the toroidal flux direction is parallel to the core of the torus (Figures 2.9 and 2.10).

The major scale is in inches.

A copper wire with a bifilar winding has *opposing magnetic fields that cancel each other.*

Take an enamel copper wire and nick the middle, then twist the wire into a double helix.

Now curl the wire into a spiral. This will generate scalar waves when there is a current. The current may change the impulses if it is relevant in tissues, thus helping the tissue if there is a therapeutic tissue need such as above a break in bone, but if it harms the tissue, it is then a stray or dirty EMF impulse.

It is believed that scalar waves have different effects, including the distortion of the temporal field. Gibbs says that the caduceus wound coil can create tachyons and alter temporal phenomena, which tends to produce waves that heal.[22]

The hyperdimensional resonator (HDR) caduceus coil built by Gibbs should have the ability to generate an electromagnetic field that disrupts time waves and alters the flow of chroniton particles.[22] The experiment with a watch subjected to the "Gibbs effect"

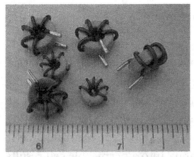 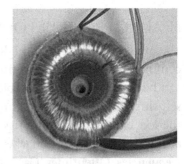

Several small toroidal inductors   A small torodial transformer

FIGURE 2.9 Several small toroidal inductors. A small toroidal transformer.

establishes that something is causing a disruption in the normal flow of time.[22] Even wind-up watches lose time when worn by some individuals who have exceptional EMF. These are a rare class of individuals who appear to have super-sensitive EMF powers.

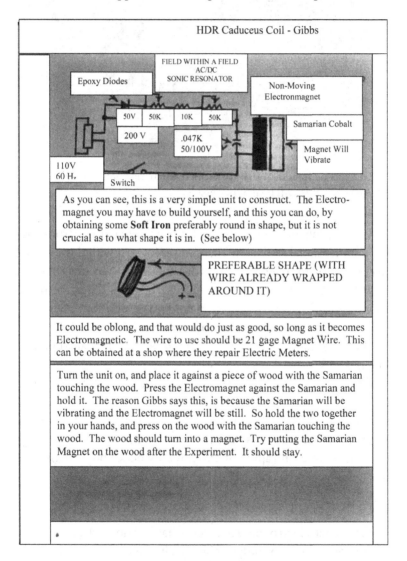

FIGURE 2.10   DR caduceus coil—Gibbs.

FIGURE 2.11   Inside of a hyperdimensional resonator (HDR) from Steven Gibbs that will heal bones faster.

This is probably caused by scalar waves emanating from the hyperdimensional resonator. These waves interact with those of the electromagnet to create a temporal disruption that can affect clocks.

According to Gibbs, this caduceus coil uses 21-gauge pure copper wires.[22] Tin alloy will mess it up. Also, aluminum wire is a poison. It generates harmonics that are wrong.

The HDR caduceus coil is the round doughnut-shaped object in Figure 2.11.

Healing properties of some frequencies, especially bones, have been developed by Becker and Marino.[23]

This configuration is also found in some other body tissues. When proper frequencies occur for treatment, they give a healing force that can even rapidly mend not only bones but soft tissues like tendons, fascia, connective tissue, and muscle.

POSITIVE EFFECTS OF ELECTROMAGNETIC FIELDS

The fact that electromagnetic fields have biological effects has been investigated not only for the health risks they invoke, but the potential for healing that has also been suggested. Much of the research was originally undertaken in those countries that had been unwilling or unable to afford the sort of expensive high-tech medical equipment used in many medical interventions in the West, though not everybody is convinced of its therapeutic effects.[11] The benefits are now being more widely explored.

How Osteoporosis Made Us Understand This Complexity

Osteoporosis was the first and best example to explain the basics of human quantum biology. Becker's[24] observations from a clinic were pretty obvious over a 10-year span. It seemed everyone had some level of the disease if it was looked for. No one was normal on their MRI or blood chemistry workups in the clinic. Was Wolff's law null and void for some reason in our modern world?

The key point of Becker's work was that for the first time in man's history, a biologic system was shown to use semiconduction as its main power source of action. Before Becker, we had two types of current conduction ideas, metallic and ionic. Metallic currents were the domain of electricians and solid-state physics and well studied. Ionic currents are based upon large ions moving current across a membrane and were well studied in cells, but they suffered from short-duration poor propagation distances in cells to be used for signal transduction, especially in the central nervous system. The third type was semiconduction, which allowed for massive changes in conduction with very small changes in physics, and it was discovered in labs in the 1930s, but few biologists understood how it worked because it works on quantum field theory, which was felt to be too ridiculously difficult to be used in cells. When Becker found bone semiconduction in the 1950s and 60s, it was a rather shocking new finding. It was so new that no one paid attention to what he found. If they had, it likely would have altered the manner in which bioenergenics in biology is studied, even up until today.

This should have stopped all biochemists dead in their tracks. *Semiconduction only occurs in materials having a very orderly molecular structure, such as crystals*, in which electrons can move easily from one electron cloud to another around one nucleus to cloud very close to one another. The atoms must be aligned in a precise geometric lattice rather than the random jumbled display that is found in most liquids and solids. Some of the crystalline materials have spaces in their lattice structures where other atoms can fit. The atoms of these "other impurities" may have more or fewer electrons than the atoms of the lattice material. Forces in this lattice structure hold the same number of electrons in place around each atom. The extra electrons of the impurities then become free to move across the crystalline structure at the speed of light, especially when very small currents are placed across this molecular arrangement. This is where grounding and light photons from the photoelectric effect affect liquid crystalline structures in cells. The number of electrons in the "impurity atom" also determines which structure will act as a positive or a negative (PN) part of the semiconductor. Semiconduction is not explained using the math of adenosine triposphate (ATP) hydrolysis or modern biochemistry.

In fact, it does not use ATP hydrolysis as its main action. This should have made biology ask better questions back then, but scientists did not. In fact, when this was published in the late 1960s, no one paid much attention to it because the biological community was not ready for the thought of semiconduction. The idea of diodes in living bone seemed completely ridiculous. Most people who heard about Becker's work ignored it because it seemed so far fetched. Becker and Bassett actually never bothered to publish follow-up experiments because of the response they got from the literature of their day.

In the mid-1960s, solid-state devices were only beginning to hit the marketplace. None of the initial designs had used continued free hole and free electron junctions' most interesting properties for industrial exploitation. When the tech industry did, it began to pass current through the positive-negative junction and found that the current made it glow. These junctions today are called LEDs, or light-emitting diodes. They are in every tech product built now.

Becker found bone worked precisely like an LED does, when no LEDs existed yet! Like any PN junction, it required an outside source of current to release its own light to generate an endogenous photoelectric effect. Researchers could not see the light, but they felt its heat in experiments. They checked on the heat generated and found that *bone gave off a blue UV light signal* when they used proper instrumentation. This is why kids with osteogenesis imperfecta have blue bone, sclera, and teeth when you see them at surgery or in a dental chair. Bone was blue colored and collagen flickered a dull brick red in their experiments. This is when they did something cool, because they found an unusual discrepancy. They combined the fluoresced light from collagen with that of apatite; they were expecting to see a fluoresced light from the whole bone, but did not. They knew right away that there had to be something else in the bone matrix to block this light for a reason. For 3 years, they could not explain it. Then the semiconductor industry found the answer in static-state electronics in a process called doping.

ELF-EMFs have been found to produce a variety of biological effects. These effects of ELF-EMFs depend upon frequency, amplitude, and length of exposure. They are also related to intrinsic susceptibility and responsiveness of different cell types. ELF-EMFs can influence cell proliferation, differentiation, cell cycle, apoptosis, DNA replication, and protein expression. This is probably why intradermal provocation and neutralization work so well in mold, pollen, food, and chemically sensitive patients. These effects are important considerations for the application of precise doses of ELF-EMFs for wound healing, tissue regeneration,[25,26] and Parkinson's and Alzheimer's diseases, as well as the hypersensitivity part of vasculitis and other inflammatory conditions.[27]

The use of a low-intensity ultrahigh-frequency (UHF) electromagnetic radiation–emitting device on sick children helped reduce the frequency of intake of anesthetics in the postoperative period, correct metabolic disorders in children with type 1 diabetes mellitus, reduce severity of diabetic nephropathy and polyneuropathy, and prevent formation of fresh foci of lipoid necrobiosis.[28]

Superimposing electromagnetic noise blocked reactive oxygen species (ROS) increase and DNA damage of the sort induced by acute exposure to 1.8 GHz RF.[29]

Bone Healing and Fracture Repair

There seems to be an increasing body of research agreeing that *pulsed electromagnetic stimulation increases bone healing*. There have been a number of theories suggesting why this should be so.

It has been suggested by Tepper[30] that magnetic fields heal fractures due to their effect on vascularity rather than osteogenesis. Pulsed electromagnetic fields were found to increase the number of bone cells, cause cell differentiation, or both.[31-34] Sun suggested it possibly resulted from the shortening of the lag phase. A further study by Sun[33] confirmed the proliferation effect on cells by pulsed EMFs (PEMFs), which also significantly altered the temporal expression of osteogenesis-related genes. Hopper[35] found that *osteoblast cells stimulated with ELF-PEMF increased endothelial proliferation 54-fold*; he believed the effect was an indirect one, altering paracrine mediators. Sollazzo[36] reported that PEMFs appeared to induce cell proliferation and differentiation by changing the action of specific genes. Noriega-Luna[37] found changes in the cytoskeletal proteins of osteoblasts following treatment with pulsed magnetic fields.

Sinusoidal EMFs (SEMFs) inhibited osteoblast proliferation, but promoted differentiation and mineralization potentials, which may help promote fracture healing.[38]

The stimulation of the formation of new blood vessels by PEMF was felt to accelerate bone fracture healing by Goto.[39] It certainly stimulates new microcirculation in the chemically sensitive patient who has lost microcirculation, causing weakness and fatigue.[39] Grana[40] and Márquez-Gamiño[41] found that short daily stimulation with pulsed electromagnetic fields accelerated bone growth and healing[42] and peri-implant bone formation, and Shen and Zhao[43] found that bone mass loss was decreased. Itoh's[44] study found that wrist fixators could be removed earlier when the fracture was stimulated with an alternating electric current.

It has been suggested that pulsed EMFs might be a new treatment for people with osteoporosis[45] due to its effects on osteoclastic-like cell formation and apoptosis. Not all studies found positive effects, and van der Jagt[46] suggested that this may be because of subtle differences in experimental setup.

Nonpulsed EMFs have also been associated with effects on bone marrow stem cells, resulting in increased differentiation into osteoblasts.[47] Yang[48] also looked at the mechanism of osteoblastic maturation following differentiation.

A review by Zhong[49] concluded that EMFs ameliorated disability due to fracture nonunion. Some patients with severe food and chemical sensitivity have difficulty stabilizing intradermal endpoint treatment so they can tolerate a substance or food. This destabilizing phenomenon probably has to do with the connective or other tissue stability, which is needed to hold the proper coherent EMF endpoint to negate the sensitivity reaction. The endpoint intradermal neutralization tool is usually very powerful for the treatment of food, mold, and chemical sensitivity.

Devices Using Electromagnetic Fields Employed in Healing

The specific devices used clinically to help stabilize and accelerate healing that help hold EMF pulses are varied, but we will discuss some as follows.

Self-Controlled Energy Neuro-Adaptive Regulation Devices

These are widely used in Russian hospitals to revitalize virtually any body system and thus promote healing. When 3000 practitioners were surveyed, they reported achieving on average:

1. 79% improvement in the musculoskeletal system; muscle injuries; and diseases such as arthritis, sciatica, lumbago, and osteoporosis

2. 82% improvement in many circulatory disorders, including strokes, thromboses, and heart failure

3. 84% improvement in virtually any respiratory problems

4. 93% improvement in both eye conditions and diseases of the digestive tract

Self-controlled energy neuro-adaptive regulation (SCENAR) analyzes the body's own electromagnetic emissions, detects any abnormalities, and automatically adjusts its output to correct the abnormality. By regularly monitoring and adjusting the immune system in this way, problems can be treated even before they are clinically detectable.

Originally developed to keep astronauts healthy during space flights, SCENAR devices are now widely used in Russian hospitals and even carried by ambulance crews because of their ability to aid recovery from cardiac arrest, accident trauma, and coma. They were also used by Russian athletes during the 2000 Australian Olympics to treat minor ailments, combat pain and fatigue, and speed up muscle repair.

Transcutaneous Electric Nerve Stimulation Machines
Transcutaneous electric nerve stimulation (TENS) is a noninvasive treatment used in physiotherapy practice to promote analgesia in acute and chronic inflammatory conditions. High-frequency (HF) and low-frequency (LF) TENS were used in a study by Sabino.[50] LF TENS had a longer-lasting effect than HF, partially due to the local release of endogenous opioids.

Aarskog[51] found that the TENS analgesic effect was dependent on whether the electrodes were placed on the same side of the body as the pain site and the stimulation intensity had to be at a subjectively strong level to work.

However, Milham reported a strong association between the use of TENS machines and other equipment such as electrical diathermy in the treatment of sports injuries and amyotrophic lateral sclerosis.

BEMER Device

A significant synergistic anti-tumor effect was found when mice were treated with a BEMER device as well as a low amount of doxorubicin. The authors of the study[52] suggested that such a combination could be useful for heavily treated patients suffering from advanced tumors and requiring additional aggressive chemotherapy that may otherwise be life threatening.

Transcranial Magnetic Stimulation and Repetitive Transcranial Magnetic Stimulation

Wassermann and Zimmermann[53] stated about transcranial magnetic stimulation (TMS): "Although we understand many of its effects at the system level, detailed knowledge of its actions, particularly as a modulator of neural activity, has lagged, due mainly to the lack of suitable non-human models". They continued, "Moderate success has been achieved in treating disorders such as depression, where the US Food and Drug Administration have cleared a TMS system for therapeutic use. In addition, there are small, but promising, bodies of data on the treatment of schizophrenic auditory hallucinations, tinnitus, anxiety disorders, neurodegenerative diseases hemiparesis, and pain syndromes."

Transcranial magnetic stimulation is rapidly developing as a powerful, noninvasive tool for studying the human brain. It can also alter the functioning of the brain beyond the time of stimulation, offering potential for therapy.[54] The effect on a reporter of exposure to TMS was an interruption of the normal exchange of information between the brain and the hand that facilitates writing, touching the nose with the hand, or raising a glass of water to the mouth to drink. The disruption stopped when the electromagnetic field was off.

The effect of 20 daily deep-TMS sessions over the prefrontal cortex of 15 patients with schizophrenia indicated improvement in cognition and negative symptoms that was maintained at 2-week post-treatment follow-up.[55]

Single or paired pulse TMS: The pulse(s) causes neurons in the neocortex under the site of stimulation to depolarize and discharge an action potential. The effect from single or paired pulses does not outlast the period of stimulation.

Repetitive TMS (rTMS): This produces effects that last longer than the period of stimulation. rTMS can increase or decrease the excitability of corticospinal or corticocortical pathways depending on the intensity of stimulation, coil orientation, and frequency of stimulation. The mechanism of these effects is not clear, although it is widely believed to reflect changes in synaptic efficacy akin to long-term potentiation and long-term depression.

It is unclear what might be the long-term effects of using a large magnetic impulse on the brain. The treatment effect could be on the whole of the brain, and it could inhibit or switch off a part.

Comparison of Treatments

Pulsed radiofrequency therapy was better at relieving low back pain than electro-acupuncture (EA) therapy, but the functional improvement of the lumbar spine was improved under EA therapy.[56]

Addictions

Transcranial magnetic therapy improved health, mood, and sleep and reduced alcohol craving in 75% of patients in the second stage of alcoholism.[57] Repeated sessions of high-frequency rTMS may be most effective in reducing the level of smoking and alcohol consumption.[58]

BIOLOGICAL RESONANCE THERAPY

Vasculitis and Chemical Sensitivity

Many chemically sensitive patients have severe muscle aches and pain, weakness, and fatigue. They have used *biological resonance therapy* for treatment. The following is our series performed by Dr. Pan and the EHC-Dallas (Table 2.6).

TABLE 2.6 50 Patients with Bioregulation Therapy (BRT)—EHC-Dallas—2015

	No. with BRT Rx	Range of Total Therapies	Average Therapies	Average Period	Effective No.	Effective (%)
Allergy	2	7	7	4 weeks	2	100
Fibromyalgia	3	1–3	2	1 week	3	100
Bell's palsy	1	6	6	1 week	1	100
Detox	3	1–3	2	2 weeks	3	100
Depression	1	1	1	1 day	1	100
EMF sensitivity	7	1–23	5	4 weeks	2	29
High triglycerides	3	1–3	2	2 weeks	3	100
Hearing loss	1	4	4	4 weeks	1	100
Tinnitus	1	7	7	7 weeks	1	100
Injury—acute	4	1–24	7	3 weeks	4	100
Hypertension	3	1–7	2	2 weeks	3	100
Vision	1	52	52	1 year	1	100
High blood sugar	2	1	1	1 day	2	100
Hypothyroidism	1	48	48	1 year	1	100
Pain	34	1–4	3	3 weeks	34	100
Kidney dysfunction	2	1–2	2	1 week	2	100
Arthritis	2	2–8	4	2 weeks	2	100
Fatigue	8	2–4	3	2 weeks	8	100
Varicosis	3	32–52	36	9 months	3	100
Insomnia	6	1–4	2	1 week	5	83

Arthritis

People with pain from osteoarthritis and rheumatoid arthritis have had significant relief from pain using therapies involving magnets, both static and "pulsed."[59-62] A pulsed electric field (PEF) was found to be effective as a potential treatment for joint pain due to cartilage degradation.[63-65]

Kumar[66] found a reduction in edema and other inflammatory effects after exposure to low-frequency pulsed electromagnetic fields.

It is important that high-strength magnets not be used all the time, as this may well reduce their effectiveness and could even have a negative effect.

Autism Spectrum Disorder

Deep rTMS may help aspects of cortical dysfunction in those with autism spectrum disorder (ASD). A new study[67] provides a potential new avenue for the development of a biomedical treatment of impaired social relating.

Bacterial and Microbial Effects

Exposing *H. pylori* to extremely low-frequency EMFs reduced its capability to protect itself.[68]

Electric fields generated using insulated electrodes can inhibit the growth of planktonic *Staphylococcus aureus* and *Pseudomonas aeruginosa*, and the effect is amplitude and frequency dependent, with a maximum at 10 MHz. The combined effect of the electric field and chloramphenicol was found to be additive.[69]

Extremely high-frequency EMF irradiation in combination with antibiotics enhanced antibacterial effects on *E. coli*.[70]

Akan[71] also found that ELF-EMFs affected bacterial growth and the response of the immune system to bacterial changes, suggesting that EMFs could be exploited for beneficial uses.

Blood Circulation

Yambe[72] used a therapeutic alternating EMF device on the hands of 11 volunteers for 20 minutes. Microcirculation changes were inferred using skin temperature measurements. The authors concluded that a significant rise in skin temperature was suggestive of a rise in peripheral circulation. This could be useful in a variety of conditions where peripheral circulation is poor. Xu[73] also found a similar result.

A pulsed low-intensity electromagnetic field (PLIMF) increased the blood flow to areas of pain or inflammation, bringing more oxygen and removing toxic substances. In a study by Durović,[74] this avoided heterotopic ossification, a complication of head and spinal cord injuries.

Ischemia-reperfusion injuries, such as those suffered from various types of cardiovascular disease, are major causes of death and disability. Studies have looked at several mechanisms of protection from the injuries caused by resumed blood flow, heat shock proteins, opioids, collateral blood flow, and nitric oxide induction, and have indicated that magnetic fields may be used as a means of providing protection via each of these mechanisms.[75]

Cancer Treatment

Pulsed EMFs reduced the blood flow (vascularization) of breast tumors,[76] thus reducing growth and metastasis.[77] Some frequencies were more effective than others, suggesting "windows" that warrant further investigation. Zimmerman[78] found that the growth of hepatocellular carcinoma (HCC) and breast cancer cells was significantly decreased by HCC- and breast cancer-specific modulation frequencies. The same frequencies did not affect proliferation of nonmalignant hepatocytes or breast epithelial cells. The authors concluded, "These findings uncover a novel mechanism controlling the growth of cancer cells at specific modulation frequencies without affecting normal tissues, which may have broad implications in oncology." The modulation frequencies are different for different cell types, and the amplitude modulation frequency range they found effective was between 100 Hz and 21 kHz, that is, the audio spectrum. So it could have implications for healing using sound (and harming using sound) as well. It is, anyway, indicative of the reality that cells (or components of cells) can detect (or demodulate) RF, despite the wireless industry and most mainline authorities repeatedly denying it. Here we have the specific amplitude modulation frequency determining the interaction. The 27 MHz RF is just the carrier.

Hu[79] found a low-level magnetic field (0.001–0.005 microteslas) was more effective than higher levels (2–5 microteslas) at limiting tumor growth. *Novikov[80] found that 0.1, 0.3, and 0.15–0.3 microteslas inhibited tumor growth*, prolonging the life of the animals involved.

Berg[81] found both pulsed electromagnetic fields and sinusoidal electromagnetic fields killed cancer cells in laboratory mice.

Short-duration electric fields were used to permanently damage cancerous tissue in aggressive cutaneous tumors in mice.[82] They affected cell membrane damage without heating and induced complete regression in 12 out of 13 treated tumors.

The application of 1500 V/cm in three sets of 10 pulses of 300 microseconds each produced the complete removal of hepatocarcinoma cells.[83] The researchers also found that multiple pulses appeared to be more effective than using one single pulse. Treatment with EMFs administered into the mouth area increased progression-free survival in patients with advanced liver cancer.[84]

Rabbits treated with alternating electric tumor treating fields (TTFs) survived longer than untreated animals.[85] This extension in survival was found to be due to an inhibition of metastatic spread, seeding, or growth in the lungs of TTF-treated rabbits compared to controls. In addition to their proven inhibitory effect on the growth of solid tumors,[86,87] TTFs may also have clinical benefit in the prevention of metastatic spread from primary tumors. *TTFs were found to slow down tumor growth in people with gliomas.*[88] Evangelou[89] found that RF-EMFs increased the number and cytotoxicity of killer cells, which they felt contributed to an improvement in end-stage cancer patients.

Plotnikov[90] found that the combination of electric fields and chemotherapy effected a significant reduction in tumor size and a prolongation of survival time. A complete cure was obtained in 33%–83% of the mice, depending on the chemotherapy used. Carefully focused microwaves were also found to increase the effectiveness of chemotherapy in treating breast cancer, shrinking tumors by nearly 50% more than chemotherapy alone.[91] Other synergistic effects were found with chemotherapy drugs and electromagnetic fields,[92] without affecting metastasis.[93]

ELF-EMFs increased the cell apoptosis effects of low doses of x-ray irradiation on liver cancer cells.[94]

Radio frequency fields in combination with single-walled carbon nanotubes (SWNTs) could produce lethal thermal injury to cancer cells. In a study by Gannon,[95] the SWNTs targeted cancer cells and left normal cells unaffected by the RF.

Some studies[96,97] have found that nanosecond pulsed electric fields (nsPEFs) could have therapeutic potential to induce apoptosis (cell death) of colon cancer cells in specified circumstances. Beebe[98] found that nsPEFs induced cell death by multiple apoptosis mechanisms that involve mitochondrial responses, potentially inducing cell death that bypasses cancer mechanisms that evade apoptosis.

The presence of more than one dental alloy in the mouth often causes pathological galvanic currents and voltage, resulting in superficial erosions of the oral mucosa and eventually in the emergence of oral cancer. Direct current electrical fields induce cancer cell death.[99]

Mi[100] also found that therapeutic radiation by steep pulsed electric field (SPEF) destroyed the integrity of induced tumor cells in mice. Their survival time was 52 days, as opposed to the 33 days of the control group.

ELF alternating magnetic fields under conditions of exposure tuned to Zn(2+) according to the ion parametric resonance (IPR) model of Blanchard and Blackman inhibited the growth of cancer and normal cells.[101]

A review of literature, mainly from Russian sources, found that low-intensity electromagnetic radiation in the millimeter band was helpful in experimental and clinical oncology. It was used in (1) preparation prior to radical treatment; (2) prevention and treatment of side effects and complications from chemotherapy and radiotherapy; (3) prevention of metastases, relapses, and dissemination of the tumor; (4) treatment of paraneoplastic syndrome; and (5) palliative therapy of incurable patients.[102]

Cancer-related frequencies appear to be tumor specific, and treatment with tumor-specific frequencies is feasible, well tolerated, and may have biological efficacy in patients with advanced cancer.[103]

ADVERSE BIOLOGICAL EFFECTS OF RADIO FREQUENCIES AND ELECTROMAGNETIC FIELDS

The adverse effects of EMF frequencies on the individual are legion because of the coherence phenomenon that is common to bacteria, viruses, parasites, molds, mycotoxin, foods, chemicals, and other physical phenomena. These will be discussed under acute and chronic categories so the clinician can determine a clinical picture that may be relevant for diagnosis and treatment. *Dirty electricity effects can be acute and/or chronic or both.*

Acute Effects

In susceptible individuals, excessive RF dirty electricity exposure can provoke acute symptoms. The most common symptoms are sleeping disturbance, headache, irritability, fatigue, short-term memory loss, and concentration difficulties. Other symptoms may include depression, dizziness, tinnitus, burning and flushed skin, digestive disturbance, tremor, and cardiac irregularities. These symptoms are often mild and transient and irritate the patient but often do not manipulate them, although some chemically sensitive and chronic degenerative disease patients will become nonfunctional with chemical and EMF exposure.

As physicians, some of us have seen patients who are experiencing this problem and are aware of the connection to RF dirty electricity exposure. *Research suggests that 3%–5% of the population fits into this category and is ever growing.*[90] If this is the case, there may be 4700 people in a town of 100,000 in the United States who react to RF exposure in some way and don't know it.

These symptoms are not uncommon in the population. In all probability, there are many other people who are having problems with insomnia or fatigue—problems provoked by dirty EMF exposure—but are unaware of the connection between cause and effect.

Any significant increase in RF exposure in our residential or work areas will make these individuals more symptomatic. Such increases are likely to push additional individuals above their tolerance threshold, producing new cases of these problems. If increased RF levels from repeated daily transmissions between smart meters and their control towers pushed an additional 1% of the community into acute reactivity to RF exposures, this would mean an additional 1500 people with insomnia, sleep disorder, central sleep apnea, headaches, fatigue, ringing in the ears, or other debilitating symptoms. These adverse responses are happening today in the general population, and most clinicians are unaware of the etiology of the problem. They certainly occur in electrically sensitive patients, causing disruption of their physiology and well-being.

Chronic Effects

Chronic exposure to RF can also cause chronic physiologic changes, including altered endocrine function (both melatonin and other hormones) and increased oxidative stress and inflammation that can lead to chronic fatigue; chemical and EMF hypersensitivity; and increased levels of cancer and male infertility, arteriosclerosis, and neurodegenerative disease, and brain dysfunction, short-term memory loss, depression, and suicidal thoughts have also become prominent. The public is already being subjected to increased levels of RF from wireless communications, cell phones and computers, and so on. Increasing the total body pollutant load of EMF and RF transmissions will further increase the occurrence of these adverse consequences.

As shown at the beginning of the last century, people began to use vehicles powered by internal combustion engines that burned gasoline. Gasoline power was cheap and convenient and greatly increased the mobility of the population. Companies sold gas and many cars were developed. Simultaneously, diseases developed and few clinicians realized the cause.

This use of fossil fuels has had long-term consequences: increased atmospheric CO_2, NO_2, CO ozone, SO_2, particulates, heavy metals, pesticides, herbicides, and other inorganic and organic chemicals with individual susceptibility to disease and individual malfunction. This greenhouse effect would lead to global climate change, which may also have adverse effects on the individual.

The use of wireless communications technology is following a similar trajectory, augmented by physicians' ignorance of the adverse effects of EMF. Wireless communication is convenient and increases our mobility. *The installation of wireless networks is also significantly initially cheaper than installation of hardwired networks.* The companies that provide these networks and the tools that are used to access them are making a great deal of money, which is great as long as the health effects are considered in the equation; mostly, they are not, for we have seen many patients made ill by wireless communications, and the price of it makes the cost much higher when broken down.

The sections of this book describe the increasing body of varied science that clearly demonstrates the existence of adverse biological effects from chronic RF exposure due to excess body pollutant load. As the *wireless communications infrastructure continues to grow, the magnitude and duration of public exposure are going to continue to increase,* and the number of people with acute or chronic effects from this exposure will continue to grow. As recognition of the problem by the public increases, exposures and infrastructure that are currently unquestioned will become politically unacceptable. However, at this time, protection against EMF- and RF-emitting towers and equipment is extremely difficult to institute and carry out. Many dirty electrical generators are well established all over the United States, making it easier for new towers to be built. The scope of this public health problem is vast, and clinicians are not aware and not being taught how to diagnose the causes and treat the results other than medication.

The axiom that has evolved for prevention and proper treatment is: don't use RF when you don't have to. Go hard-wired wherever it is feasible to do so. If you do use RF, design the technology to use as little of it as possible. This is extremely difficult at this time because so many of the emitters are beyond the individual's control in the areas where broadcasting is commonplace.

Current engineering choices in advanced meter (AMI) technology have not been designed with these goals in mind, since the industry has not had a practical incentive to recognize the problem and to "work the problem." But as a purchaser of technology, one could choose to push vendors toward designing and providing hardware options that would address these goals. This would put the public in the position of being part of the solution rather than just another part of the problem. Manufacturers often listen, but it takes time to institute new ideas and procedures. A great example is the construction of less-polluted buildings for homes and public buildings. The knowledge is there, but only a few have been constructed over the last 30 years.

Opt-Out Program

It has been suggested that people who have problems with EHS or concerns about health exposures to RF can be taken care of by creating an "opt-out" program, allowing them to decline the installation of a smart meter on their home. These suggestions, though better for the individual, overlook some obvious and important problems:

You can't opt out of exposure to your neighbor's meter, which is 10 feet away from your bedroom window. Or all the meters are on the wall of the rental apartment complex: the frequencies come right in. Or the ones on the wall of the complex right across the alley from your apartment; the same happens. You can't opt out of exposure to the meter on the other side of your bedroom wall if you are a baby in a crib or of exposure to transmissions from the radio (EMF) tower 100 meters from your house.

The idea of an opt-out program is an effort to address the concerns of people who are personally worried about RF exposures, either because they are aware of having acute reactions to these exposures or because they have a general concern about the acute or chronic effects from such exposure.

A voluntary opt-out program does not protect the community at large from adverse effects that they are unaware of and unconcerned about. For example, the current research shows that cancer rates are higher in residences near cellular transmission towers.[104] How does a voluntary opt-out program help the person who develops breast cancer 3 years after installation of a transmission tower across the street from her house? She didn't know it was a problem. What about the individual who has cardiovascular disease? Will opting out help for the same reasons, or will sudden exposure from another source cause arrhythmia and death? Can it help with the individual who develops neurodegenerative disease and can't function efficiently when the cause, the source, is beamed at him/ her? Pretty soon, a whole portion of the population needs to opt out, and this would cause havoc in these areas. Therefore, other shielding solutions must be found at the source. Furthermore, wire connections or some other not-yet-developed technology will need to be found and used. The wireless network will now become the biggest epidemic man has ever seen.

The Body's Mechanism of Handling Electromagnetic Fields

The sensory nervous system is in the vascular tree, skin, olfactory nerves, respiratory tract, and GI or GU tract and will serve as an intake apparatus for electromagnetic waves. When sensitized by EMF waves or chemicals, this system then yields hyper-responses that can produce adverse sensitivity and signs in the body. At times, these EMF frequencies affect the body so severely that they render it nonfunctional, along with these sensitized nerves, which are usually those of the connected tissue (CT) matrix.

The extracellular matrix or the ground regulation system (GRS) is the main environmental receptor system of the body.[105] This system can amplify the EMF impulse, distributing its adverse reaction throughout the body. The GRS is an open-ended dynamic molecular dissipative energy system in the body that is labile because of its molecular oscillation and the fact it is a receptor for all environmental stimuli, including electromagnetic, electric, atmospheric pressure, static electricity, spherics, circadian rhythms and cycles, nutrition and total body pollutant or traumatic load, life forces, and subtle energy.[105]

The ground regulation system controls homeostasis and allostasis and is part of every defense and inflammatory reaction. The regulation of this system depends upon the spontaneity of molecular reactions, with homeostasis being a dynamic balance between *entropy* (the random distribution of molecules) and *enthalpy* (the structure organization of molecules). Excess entropy means that there is an excess of exchanges of energy and thus structure is lost; therefore, oxidative stress, acute inflammation, allergy, rheumatoid disease, and tumors occur. *Excess enthalpy means there is too little free exchange of energy; therefore, supermolecular states of order occur with sclerosis, nodules, sarcomas, and aneurysms,* emphasizing the idea that energy ($E = mc^2$) can be changed into mass and that inappropriate input of pollutants, bacteria, or viral energy results in pathologic processes.[105] This receptor system of the body receives all impulses and this certainly applies to EMF.

According to Panagopoulas and Margaritis, both systems of digital mobile telephony radiation used in Europe, GSM 900 MHz and DCS 1800 MHz, and the system used in the United States, GSM 1900 MHz, use different carrier frequencies (900, 1800, and 1900 MHz, respectively) but the same pulse repetition frequency of 217 Hz.[106] As is obvious, the signals of digital mobile telephony radiation combine radio frequencies and extremely low frequencies. All three systems use the time division multiple access (TDMA) code to increase the number of people who can simultaneously communicate with a base station. Radiation is emitted in frames of 4.615 msec duration, at a repetition rate of 217 Hz. Each frame consists of eight time slots and each user occupies one of them. Within each time slot, the microwave radiation uses a type of phase modulation (Gaussian minimum shift keying modulation [GMSK]) to carry the information. The transmitted frames by both handsets and base stations are grouped into multiframes of 25 by the absence of every 26th frame. This results in an additional multiframe repetition frequency of 8.34 Hz. Finally, handsets emit an even lower frequency at 2 Hz whenever the user is not speaking, for energy-saving reasons (nonmodulated or nonspeaking emission or discontinuous transmission mode [DTX]). Of course, when the handsets operate in DTX mode, the average emitted power is much less (about one-tenth of the emitted power when they operate in speaking mode).[107,108]

Except for the carrier frequency, another important difference between the three systems of digital mobile telephony radiation is that GSM 900 MHz antennas of both mobile phones and base stations operate with double the output power of the corresponding DCS 1800 MHz or GSM 1900 MHz ones. GSM 900 MHz handsets operate with 2 W peak power output, while DCS 1800 and GSM 1900 MHz ones operate with 1 W peak power output.

There appear to be basic emanations from the generation of electricity. There are (1) electromagnetic effects from fixed power sources, supplies, and specialized equipment; (2) electromagnetic smog (dirty electricity) from communication equipment from home EMF, computers, Wi-Fi, smart meters, and so on; and (3) ground currents that they generate. Each will be discussed individually. These three emanations make up the dirty electricity that disturbs membrane and internal physiologic mechanisms that when disrupted long-term can lead to chronic pathology, resulting in weakness, fatigue, end organ failure, and death. The total environmental pollutant load when chronically expressed can lead to the existence of the total body pollutant load and disease.

REFERENCES

1. What Is Schumann Resonance? Archive of Dr. Magneto's Questions and Answers. *NASA*. Retrieved 2015-11-23. http://image.gsfc.nasa.gov/poetry/ask/q768.html.
2. What Are Electromagnetic Fields? 2015. *WHO*. Retrieved 2015-11-23. http://www.who.int/peh-emf/about/WhatisEMF/en/.
3. Chapter 2—Health and EMF. 1996. *Parliament of Australia*. Retrieved 2015-11-23. http://www.aph.gov.au/Parliamentary_Business/Committees/Senate/Economics/Completed_inquiries/pre1996/elec/report/c02.
4. Goldsworthy, A. 2007. The biological effects of weak electromagnetic fields. Retrieved 2015-11-23. http://apps.fcc.gov/ecfs/document/view?id=7521098551.
5. Marshall, B. 2000. How Radio Works. HowStuffWorks.com. Retrieved 2015-11-23. https://www.google.com/search?source=hp&ei=vo5gW_CtAtLysQWXm6PYBQ&q=Marshall%2C+B.+2000+How+Radio+Works.+Hot+Stuff+Works.com.
6. Curtis, T. S. 1916. *High Frequency Apparatus: Its Construction and Practical Application*. New York ,USA: Everyday Mechanics Company. p. 6.
7. Mieny, C. J. 2003. *Principles of Surgical Patient Care* (2nd ed.). New Africa Books. p. 136.
8. Herschel, J. F. W. 1847. *Results of Astronomical Observations Made during the Years 1834,5,6,7,8 at the Cape of Good Hope; Being the Completion of a Telescopic Survey of the Whole Surface of the Visible Heavens, Commenced in 1825*. London: Smith, Elder and Co.
9. Ritter, J. W. 1801. Versuche und Bermerkungen über den Galvanismus der Voltaischen Batterie. *Annalen der Physik*. 7: 431–484.
10. Faraday, M. 1933. *Faraday's Diary*. Volume IV, Nov. 12, 1839–June 26, 1847 (Thomas Martin, ed.). London: George Bell and Sons, Ltd.
11. Maxwell, J. C. 1873. *Year 13–1873: A Treatise on Electricity and Magnetism by James Clerk Maxwell*. MIT Libraries.
12. Hertz, H. 1893. Electric Waves; Being Researches on the Propagation of Electric Action with Finite Velocity through Space. Cornell University Library Historical Monographs Collection. Reprinted by Cornell University Library Digital Collections.
13. Röntgen, W. C. (1845–1923). Royal Netherlands Academy of Arts and Sciences. Retrieved 2015-7-20. http://www.interacademies.org/Netherlands.aspx.

14. Villard, P. 1900. Sur la réflexion et la réfraction des rayons cathodiques et des rayons déviables du radium (On reflection and refraction of cathodiqus rays and radium rays devables). *Comptes rendus.* 130: 1010–1012.
15. Andrade, E. N. da C. 1962. Some Personal Reminiscences. *Fifty Years of X-Ray Diffraction.* Springer, 507–513.
16. Smith, C. W., and Monro, J. 1988. Electromagnetic effects in humans. In: Fröhlich, H. (ed.), *Biological Coherence and Response to External Stimuli.* Berlin: Springer-Verlag, pp. 205–232.
17. Rea, W., M. D. FACS, Electromagnetic field sensitivity. *J. Bioelectr.* 10(1&2): 241–256.3.
18. Schumann I. 1979. Preoperative measures to promote wound healing. *Nurs. Clin. North Am.* 14(4): 683.
19. Hardell, L., and C. Sage. 2008. Biological effects from electromagnetic field exposure and public exposure standards. *Biomed. Pharmacother.* 62(2): 104–109.
20. Belpomme, D., C. Campagnac, and P. Irigaray. 2015. Reliable disease biomarkers characterizing and identifying electrohypersensitivity and multiple chemical sensitivity as two etiopathogenic aspects of a unique pathological disorder. *Rev. Environ. Health.* 30(4): 251–271.
21. Smith, C. 2010. The essential unity of CAM. *J. Altern. Complement. Med.* 16(9): 931–933.
22. Gibbs, S. HDR Caduceus Coil. Illustration by Steven Gibbs. Retrieved 2015-11-23. http://fluxcap.com/HDR_Caduceus_Coil. htm.
23. Marino, A., and R. Becker. 1977. Biological effects of extremely low frequency electric and magnetic fields: A review. *Physiol. Chem. Phys.* 9: 131.
24. Becker, R. 1985. *The Body Electric: Electromagnetism and the Foundation of Life.* New York: William Morrow & Co., Inc.
25. Hug, K., and M. Röösli. 2011. Therapeutic effects of whole-body devices applying pulsed electromagnetic fields (PEMF): A systematic literature review. *Bioelectromagnetics* 33(2): 95–105. doi: 10.1002/bem.20703.
26. Patruno, A., P. Amerio, M. Pesce, G. Vianale, S. Di Luzio, A. Tulli, S. Franceschelli, A. Grilli, R. Muraro, and M. Reale. 2010. Extremely low frequency electromagnetic fields modulate expression of inducible nitric oxide synthase, endothelial nitric oxide synthase and cyclooxygenase-2 in the human keratinocyte cell line HaCat: Potential therapeutic effects in wound healing. *Br. J. Dermatol.* 162(2): 258–266.

27. Patruno, A., S. Tabrez, P. Amerio, M. Pesce, G. Vianale, S. Franceschelli, A. Grilli, M. A. Kamal, and M. Reale. 2012. Kinetic study on the effects of extremely low frequency electromagnetic field on catalase, cytochrome P450 and inducible nitric oxide synthase in human HaCaT and THP-1 cell lines. *CNS & Neurol. Disord.- Drug. Targets CNSNDDT.* 936–944.

28. Azov, N. A., and E. A. Azova. 2009. Application of low-intensity and ultrahigh frequency electromagnetic radiation in modern pediatric practice. *PUBMed* (5): 34–37.

29. Yao, K., W. Wu, K. Wang, S. Ni, P. Ye, Y. Yu, J. Ye, and L. Sun. 2008. Electromagnetic noise inhibits radiofrequency radiation-induced DNA damage and reactive oxygen species increase in human lens epithelial cells. *Mol. Vis.* 14: 964–969.

30. Tepper, O. M. 2004. Electromagnetic fields increase *in vitro* and *in vivo* angiogenesis through endothelial release of FGF-2. *FASEB J*(11): 1231–1233.

31. Chang, W. H.-S., L.-T. Chen, J.-S. Sun, and F.-H. Lin. 2004. Effect of pulse-burst electromagnetic field stimulation on osteoblast cell activities. *Bioelectromagnetics.* 25(6): 457–465.

32. Tsai, M.-T., W.-J. Li, R. S. Tuan, and W. H. Chang. 2009. Modulation of osteogenesis in human mesenchymal stem cells by specific pulsed electromagnetic field stimulation. *J. Orthop. Res.* 27(9): 1169–1174.

33. Sun, L.-Y., D.-K. Hsieh, P.-C. Lin, H.-T. Chiu, and T. W. Chiou. 2010. Pulsed electromagnetic fields accelerate proliferation and osteogenic gene expression in human bone marrow mesenchymal stem cells during osteogenic differentiation. *Bioelectromagnetics.* 31(3): 209–219.

34. Feng, X., X. He, K. Li, W. Wu, X. Liu, and L. Li. 2011. The effects of pulsed electromagnetic fields on the induction of rat bone marrow mesenchymal stem cells to differentiate into cardiomyocytes-like cells *in vitro*. *J. Biomed. Eng.* 28(4): 676–682.

35. Hopper, R. A., J. P. Verhalen, O. Tepper, B. J. Mehrara, R. Detch, E. I. Chang, S. Baharestani, B. J. Simon, and G. C. Gurtner. 2009. Osteoblasts stimulated with pulsed electromagnetic fields increase HUVEC proliferation via a VEGF-A independent mechanism. *Bioelectromagnetics.* 30(3): 189–197.

36. Sollazzo, V., A. Palmieri, F. Pezzetti, L. Massari, and F. Carinci. 2010. Effects of pulsed electromagnetic fields on human osteoblastlike cells (MG-63): A pilot study. *Clin. Orthop. Relat. Res.* 468(8): 2260–2277.

37. Noriega-Luna, B., M. Sabanero, M. Sosa, and M. Avila-Rodriguez. 2011. Influence of pulsed magnetic fields on the morphology of bone cells in early stages of growth. *Micron.* 42(6): 600–607.

38. Zhou, J., L. Ming, B. Ge, J. Wang, R. Zhu, Z. Wei, H. Ma, C. J. Xian, and K. Chen. 2011. Effects of 50 Hz sinusoidal electromagnetic fields of different intensities on proliferation, differentiation and mineralization potentials of rat osteoblasts. *Bone.* 49(4): 753–761.

39. Goto, T., M. Fujioka, M. Ishida, M. Kuribayashi, K. Ueshima, and T. Kubo. 2010. Noninvasive up-regulation of Angiopoietin-2 and Fibroblast Growth Factor-2 in bone marrow by pulsed electromagnetic field therapy. *J. Orthop. Sci.* 15(5): 661–665.

40. Grana, D. R., H. J. Marcos, and G. A. Kokubu. 2008. Pulsed electromagnetic fields as adjuvant therapy in bone healing and peri-implant bone formation: An experimental study in rats. *Acta. Odontol. Latinoam.* 21(1): 77–83.

41. Márquez-Gamiño, S., F. Sotelo, M. Sosa, C. Caudillo, G. Holguín, M. Ramos, F. Mesa, J. Bernal, and T. Córdova. 2008. Pulsed electromagnetic fields induced femoral metaphyseal bone thickness changes in the rat. *Bioelectromagnetics.* 29(5): 406–409.

42. Gupta, A., K. Srivastava, and S. Avasthi. 2009. Pulsed electromagnetic stimulation in nonunion of tibial diaphyseal fractures. *Indian. J. Orthop.* 42(2): 156–160.

43. Shen, W., and J. Zhao. 2010. Pulsed electromagnetic fields stimulation affects BMD and local factor production of rats with disuse osteoporosis. *Bioelectromagnetics.* 31(2): 113–119.

44. Itoh, S., T. Ohta, Y. Sekino, Y. Yukawa, and K. Shinomiya. 2008. Treatment of distal radius fractures with a wrist-bridging external fixation: The value of alternating electric current stimulation. *J. Hand Surgery (European Volume).* 33(5): 605–608.

45. Bai, M. H., B. F. Ge, Z. Wei, J. Bai, and Z. F. Cheng. 2009. Effect of pulsed electromagnetic field on the changes of osteoclasts in ovariectomized rats bone marrow culture *in vitro. China J. Orthop. Traumatol.* 22(10): 727–729.

46. Van Der Jagt, O. P. J. C. Van Der Linden, J. H. Waarsing, J. A. N. Verhaar, and H. Weinans. 2012. Systemic treatment with pulsed electromagnetic fields do not affect bone microarchitecture in osteoporotic rats. *Int. Orthop. (SICOT).* 36(7): 1501–1506.

47. Yang, Y., C. Tao, D. Zhao, F. Li, W. Zhao, and H. Wu. 2010. EMF acts on rat bone marrow mesenchymal stem cells to promote differentiation to osteoblasts and to inhibit differentiation to adipocytes. *Bioelectromagnetics.* 31(4): 277–285.

48. Yang, J.-C., S. Lee, C. Chen, C. Lin, C. Chen, and H. Huang. 2010. The role of the calmodulin-dependent pathway in static magnetic field-induced mechanotransduction. *Bioelectromagnetics.* 31(4): 255–261.

49. Zhong, C., T. F. Zhao, Z. J. Xu, and R. X. He. 2012. Effects of electromagnetic fields on bone regeneration in experimental and clinical studies: A review of the literature. *Chin. Med. J. (Engl).* 125(2): 367–372.

50. Sabino, G. S., C. M. F. Santos, J. N. Francischi, and M. Antônio De Resende. 2008. Release of endogenous opioids following transcutaneous electric nerve stimulation in an experimental model of acute inflammatory pain. *J. Pain.* 8(2): 157–163.

51. Aarskog, R., M. I. Johnson, J. Hendrik Demmink, A. Lofthus, V. Iversen, R. Lopes-Martins, J. Joensen, and J. M. Bjordal. 2007. Is mechanical pain threshold after transcutaneous electrical nerve stimulation (TENS) increased locally and unilaterally? A randomized placebo-controlled trial in healthy subjects. *Physiother. Res. Int.* 12(4): 251–263.

52. Říhová, B., T. Etrych, M. Šírová, J. Tomala, K. Ulbrich, and M. Kovář. 2011. Synergistic effect of EMF–BEMER-type pulsed weak electromagnetic field and HPMA-bound doxorubicin on mouse EL4 T-cell lymphoma. *J. Drug. Target.* 19(10): 890–899.

53. Wassermann, E. M., and T. Zimmermann. 2012. Transcranial magnetic brain stimulation: Therapeutic promises and scientific gaps. *Pharmacol. Ther.* 133(1): 98–107.

54. Hallett, M. 2000. Transcranial magnetic stimulation and the human brain. *Nature.* 406(6792): 147–150.

55. Levkovitz, Y., L. Rabany, E. Vadim Harel, and A. Zangen. 2011. Deep transcranial magnetic stimulation add-on for treatment of negative symptoms and cognitive deficits of schizophrenia: A feasibility study. *Int. J. Neuropsychopharmacol.* 14(7): 991–996.

56. Lin, M., M. Lin, J. Fen, W. Lin, C. Lin, and P. Chen. 2010. A comparison between pulsed radiofrequency and electro-acupuncture for relieving pain in patients with chronic low back pain. *Acupunct. Electrother. Res.* 35(3–4): 133–146.

57. Staroverov, A. T., V. B. Vil'Yanov, Yu. M. Raigorodskii, and M. A. Rogozina. 2008. Transcranial magnetotherapy in the complex treatment of affective disorders in patients with alcoholism. *Neurosci. Behav. Physi.* 108(4): 507–511.

58. Barr, M. S., F. Farzan, V. C. Wing, T. P. George, P. B. Fitzgerald, and Z. J. Daskalakis. 2011. Repetitive transcranial magnetic stimulation and drug addiction. *Int. Rev. Psychiatry.* 23(5): 454–466.

59. Pipitone, N., and D. L. Scott. 2001. Magnetic pulse treatment for knee osteoarthritis: A randomised, double-blind, placebo-controlled study. *Curr. Med. Res. Opin.* 17(3): 190–196.
60. Hinman, M. R., J. Ford, and H. Heyl. 2002. Effects of static magnets on chronic knee pain and physical function: A double-blind study. *Altern. Ther. Health Med.* 8(4): 50–55.
61. Wolkso, P. M., D. M. Eisenberg, L. S. Simon, R. B. Davis, J. Walleczek, M. Mayo-Smith, T. J. Kaptchuk, and R. S. Phillips. 2004. Double-blind placebo-controlled trial of static magnets for the treatment of osteoarthritis of the Knee: Results of a pilot study. *Altern. Ther. Health Med.* 10(2): 36–43.
62. Ganesan, K., and A. C. Gengadharan. 2009. Low frequency pulsed electromagnetic field—a viable alternative therapy for arthritis. *Indian J. Exp. Biol.* 47(12): 939–948.
63. Fitzsimmons, R. J., S. L. Gordon, J. Kronberg, T. Ganey, and A. A. Pilla. 2008. A pulsing electric field (PEF) increases human chondrocyte proliferation through a transduction pathway involving nitric oxide signaling. *J. Orthop. Res.* 26(6): 854–859.
64. Luo, Q., S. Li, C. He, H. He, L. Yang, and L. Deng. 2008. Pulse electromagnetic fields effects on Serum E2 levels, chondrocyte apoptosis, and matrix metalloproteinase-13 expression in ovariectomized rats. *Rheumatol. Int.* 29(8): 927–935.
65. Ongaro, A., K. Varani, F. F. Masieri et al. 2012. Electromagnetic fields (EMFs) and adenosine receptors modulate prostaglandin E 2 and cytokine release in human osteoarthritic synovial fibroblasts. *J. Cell. Physiol.* 227(6): 2461–2469.
66. Kumar, V. S., D. A. Kumar, K. Kalaivani, A. C. Gangadharan, K. V. S. Narayana Raju, P. Thejomoorthy, B. M. Manohar, and R. Puvanakrishnan. 2005. Optimization of pulsed electromagnetic field therapy for management of arthritis in rats. *Bioelectromagnetics.* 26(6): 431–439.
67. Enticott, P. G., H. A. Kennedy, A. Zangen, and P. B. Fitzgerald. 2011. Deep repetitive transcranial magnetic stimulation associated with improved social functioning in a young woman with an autism spectrum disorder. *J. ECT.* 27(1): 41–43.
68. Di Campli, E., S. Di Bartolomeo, R. Grande, M. Di Giulio, and L. Cellini. 2009. Effects of extremely low-frequency electromagnetic fields on *Helicobacter pylori* biofilm. *Curr. Microbiol.* 60(6): 412–418.
69. Giladi, M., Y. Porat, A. Blatt, Y. Wasserman, E. D. Kirson, E. Dekel, and Y. Palti. 2008. Microbial growth inhibition by alternating

electric fields. *Antimicrob. Agents And Chemother.* 52(10): 3517–3522.

70. Torgomyan, H., H. Tadevosyan, and A. Trchounian. 2010. Extremely high frequency electromagnetic irradiation in combination with antibiotics enhances antibacterial effects on *Escherichia coli. Curr. Microbiol.* 62(3): 962–967.

71. Akan, Z., B. Aksu, A. Tulunay, S. Bilsel, and A. Inhan-Garip. 2010. Extremely low-frequency electromagnetic fields affect the immune response of monocyte-derived macrophages to pathogens. *Bioelectromagnetics.* 31(8): 603–612.

72. Yambe, T., A. Inoue, K. Sekine, Y. Shiraishi, M. Watanabe, T. Yamaguchi, M. Shibata, M. Maruyama, S. Konno, and S. Nitta. 2005. Effect of the alternative magnetic stimulation on peripheral circulation for regenerative medicine. *Biomed. Pharmacother.* 59 (Suppl 1): S174–176.

73. Xu, S. 2001. Acute effects of whole-body exposure to static magnetic fields and 50-Hz electromagnetic fields on muscle microcirculation in anesthetized mice. *Bioelectrochemistry.* 53(1): 127–135.

74. Djurovic, A., D. Miljkovic, Z. Brdareski, A. Plavsic, and M. Jevtic. 2009. Pulse low-intensity electromagnetic field as prophylaxis of heterotopic ossification in patients with traumatic spinal cord injury. *Vojnosanitetski Pregled VSP.* 66(1): 22–28.

75. Robertson, J. A., A. W. Thomas, Y. Bureau, and F. S. Prato. 2007. The influence of extremely low frequency magnetic fields on cytoprotection and repair. *Bioelectromagnetics.* 28(1): 16–30.

76. Williams, C. D., and M. S. Markov. 2001. Therapeutic electromagnetic field effects on angiogenesis during tumor growth: A pilot study in mice. *Electro. Magnetobiol.* 21(6A): 323–329.

77. Cameron, I. L., L. Z. Sun, N. Short, W. E. Hardman, and C. D. Williams. 2005. Therapeutic electromagnetic field (TEMF) and gamma irradiation on human breast cancer xenograft growth, angiogenesis and metastasis. *Cancer Cell Int.* 5: 23.

78. Zimmerman, J. W., M. J. Pennison, I. Brezovich et al. 2011. Cancer cell proliferation is inhibited by specific modulation frequencies. *Br. J. Cancer.* 106(2): 307–313.

79. Hu, J. H., L. S. St-Pierre, C. A. Buckner, R. M. Lafrenie, and M. A. Persinger. 2010. Growth of injected melanoma cells is suppressed by whole body exposure to specific spatial-temporal configurations of weak intensity magnetic fields. *Int. J. Radiat. Biol.* 86(2): 79–88.

80. Novikov, G. V., V. V. Novikov, and E. E. Fesenko. 2009. Effect of weak combined static and low-frequency alternating magnetic fields on the Ehrlich ascites carcinoma in mice. *Biophysics.* 54(6): 741–747.
81. Berg, H., B. Günther, I. Hilger, M. Radeva, N. Traitcheva, and L. Wollweber. 2010. Bioelectromagnetic field effects on cancer cells and mice tumors. *Electromagn. Biol. Med.* 29(4): 132–143.
82. Al-Sakere, B., F. Andre, C. Bernat, E. Connault, P. Opolon, R. V. Davalos, B. Rubinsky, and L. M. Mir. 2007. Tumor ablation with irreversible electroporation. *PLOS ONE.* 2(11): e1135.
83. Miller, L., J. Leor, and B. Rubinsky. 2005. Cancer cells ablation with irreversible electroporation. *Technol. Cancer Res. Treat.* 4(6): 699–705.
84. Costa, F. P., A. C. De Oliveira, R. Meirelles et al. 2011. Treatment of advanced hepatocellular carcinoma with very low levels of amplitude-modulated electromagnetic fields. *Br. J. Cancer.* 105(5): 640–648.
85. Kirson, E. D., M. Giladi, Z. Gurvich, A. Itzhaki, D. Mordechovich, R. S. Schneiderman, Y. Wasserman, B. Ryffel, D. Goldsher, and Y. Palti. 2009. Alternating electric fields (TTFields) inhibit metastatic spread of solid tumors to the lungs. *Clin. Exp. Metastasis.* 26(7): 633–640.
86. Kirson, E. D. 2004. Disruption of cancer cell Replication by alternating electric fields. *Cancer Res.* 64(9): 3288–3295.
87. Salzberg, M., E. Kirson, Y. Palti, and C. Rochlitz. 2008. A pilot study with very low-intensity, intermediate-frequency electric fields in patients with locally advanced and/or metastatic solid tumors. *Onkologie.* 31(7): 362–365.
88. Kirson, E. D., V. Dbaly, F. Tovarys et al. 2007. Alternating electric fields arrest cell proliferation in animal tumor models and human brain tumors. *Proc. Natl. Acad. Sci.* 104(24): 10152–10157.
89. Evangelou, A., I. Toliopoulos, C. Giotis, A. Metsios, I. Verginadis, Y. Simos, K. Havelas, G. Hadziaivazis, and S. Karkabounas. 2011. Functionality of natural killer cells from end-stage cancer patients exposed to coherent electromagnetic fields. *Electromagn. Biol. Med.* 30(1): 46–56.
90. Plotnikov, A., D. Fishman, T. Tichler, R. Korenstein, and Y. Keisari. 2004. Low electric field enhanced chemotherapy can cure mice with CT-26 colon carcinoma and induce anti-tumour immunity. *Clin. Exp. Immunol.* 138(3): 410–416.

91. Dooley, W. C., H. I. Vargas, A. J. Fenn, M.-B. Tomaselli, and J. K. Harness. 2009. Focused microwave thermotherapy for preoperative treatment of invasive breast cancer: A review of clinical studies. *Ann. Surg. Oncol.* 17(4): 1076–1093.

92. Tofani, S. 2003. Static and ELF magnetic fields enhance the *in vivo* anti-tumor efficacy of Cis-platin against Lewis Lung Carcinoma, but not of cyclophosphamide against B16 melanotic melanoma. *Pharmacol. Res.* 48(1): 83–90.

93. Zhang, D., X. Pan, S. Ohno, T. Osuga, S. Sawada, and K. Sato. 2011. No effects of pulsed electromagnetic fields on expression of cell adhesion molecules (integrin, CD44) and matrix metalloproteinase-2/9 in osteosarcoma cell lines. *Bioelectromagnetics.* 32(6): 463–473.

94. Jian, W., Z. Wei, C. Zhiqiang, and F. Zheng. 2009. X-ray-induced apoptosis of BEL-7402 cell line enhanced by extremely low frequency electromagnetic field *in vitro. Bioelectromagnetics.* 30(2): 163–165.

95. Gannon, C. J., P. Cherukuri, B. I. Yakobson et al. 2007. Carbon nanotube-enhanced thermal destruction of cancer cells in a noninvasive radiofrequency field. *Cancer.* 110(12): 2654–2665.

96. Hall, E. H., K. H. Schoenbach, and S. J. Beebe. 2007. Nanosecond pulsed electric fields induce apoptosis in P53-wildtype and P53-null HCT116 colon carcinoma cells. *Apoptosis.* 12(9): 1721–1731.

97. Ren, W., and S. J. Beebe. 2011. An apoptosis targeted stimulus with nanosecond pulsed electric fields (nsPEFs) in E4 squamous cell carcinoma. *Apoptosis.* 16(4): 382–393.

98. Beebe, S. J., W. E. Ford, W. Ren, X. Chen, and K. H. Schoenbach. 2009. Non-ionizing radiation with nanosecond pulsed electric fields as a cancer treatment: *In vitro* studies. *2009 Annual International Conference of the IEEE Engineering in Medicine and Biology Society* pp. 6509–6512.

99. Wartenberg, M., N. Wirtz, A. Grob, W. Niedermeier, J. Hescheler, S. C. Peters, and H. Sauer. 2008. Direct current electrical fields induce apoptosis in oral mucosa cancer cells by NADPH oxidase-derived reactive oxygen species. *Bioelectromagnetics.* 29(1): 47–54.

100. Mi, Y., C. Sun, C. Yao, L. Xiong, R. Liao, Y. Hu, and L. Hu. 2004. Lethal and inhibitory effects of steep pulsed electric field on tumor-bearing BALB/c mice. *The 26th Annual International Conference of the IEEE Engineering in Medicine and Biology Society* 7: 5005–5008.

101. Sarimov, R., E. Markova, F. Johansson, D. Jenssen, and I. Belyaev. 2005. Exposure to ELF magnetic field tuned to Zn inhibits growth of cancer cells. *Bioelectromagnetics.* 26(8): 631–638.

102. Teppone, M., and R. Avakyan. 2010. Extremely high-frequency therapy in oncology. *J. Altern. Complement. Med.* 16(11): 1211–1216.
103. Barbault, A., F. P. Costa, B. Bottger, R. F. Munden, F. Bomholt, N. Kuster, and B. Pasche. 2009. Amplitude-modulated electromagnetic fields for the treatment of cancer: Discovery of tumor-specific frequencies and assessment of a novel therapeutic approach. *J. Exp. Clin. Cancer Res.* 28: 51.
104. Dart, P., K. Cordes, A. Elliott, J. Knackstedt, J. Morgan, P. Wible, and S. Baker. 2013. Biological and health effects of microwave radio frequency transmissions. *A Review of the Research Literature.*
105. Rea, W. J., and K. Patel. 2010. *Reversibility of Chronic Degenerative Disease and Hypersensitivity.* Vol. 1. Boca Raton, FL: CRC.
106. Panagopoulos, D. J., N. Messini, A. Karabarbounis, A. L. Philippetis, L. H. Margaritis. 2000a. Radio frequency electromagnetic radiation within safety levels, alters the physiological function of insects. In: Kostarakis, P., P. Stavroulakis (eds.), *Millennium International Workshop on Biological Effects of Electromagnetic Fields, Proceedings.* pp. 169–175.
107. Harper, A. C., R. V. Buress, D. J. Panagopoulos, and L. H. Margaritis. 2008. *Mobile Telephones: Networks, Applications, and Performance.* New York: Nova Science, p. 109.
108. Panagopoulos, D. J., A. Karabarbounis, and L. H. Margaritis. 2004. Effect of GSM 900-MHz mobile phone radiation on the reproductive capacity of *Drosophila melanogaster. Electromagn. Biol. Med.* 23(1): 29–43.

Index

Printed in the United States
by Baker & Taylor Publisher Services